Eduardo Pagel Floriano

Forest Management

Eduardo Pagel Floriano

Forest Management

For sustainability and excellence

ScienciaScripts

Cover image: www.ingimage.com

This book is a translation from the original published under ISBN 978-613-9-68709-1.

Publisher:
Sciencia Scripts
is a trademark of
Dodo Books Indian Ocean Ltd. and OmniScriptum S.R.L publishing group

120 High Road, East Finchley, London, N2 9ED, United Kingdom
Str. Armeneasca 28/1, office 1, Chisinau MD-2012, Republic of Moldova, Europe
Printed at: see last page
ISBN: 978-620-8-31568-9

Eduardo Pagel Floriano

FOREST MANAGEMENT
for sustainability and excellence

Rio Largo, AL, Brazil

2024

F635
m

Floriano, Eduardo Pagel

Forest management: towards sustainability and
excellence / Eduardo
Pagel Floriano. - Rio Largo, 2024.
421 p. ; ill.

Includes bibliography and annexes.

1. Forestry production - Planning. 2. Production chains.
3.
Forestry economics. 4. forestry modeling. I. Floriano,
Eduardo Pagel. II. Title.

CDU: 630*2

PREFACE

The motivation for producing this compendium came from the need for a text to help teach forest management to future forest engineers. The few texts in Portuguese generally lack the necessary scope to serve as a basis for the subject today.

Forest management has become an extremely broad and complex subject since the certification programs, encompassing not only the forest, but practically all the resources used for forest production, from people, infrastructure, finances and the environment, involving various technologies from different sub-areas with the aim of subsidizing the production of tangible and intangible forest goods in a sustainable manner and with excellence. Forest Engineers usually specialize in one or a few areas related to forest management, due to the scope and complexity of the subject, having to call on professionals from other areas, such as biology, geography, mathematics, economics, among others, to solve the problems that arise when managing forests.

Here, we have tried to provide an introduction to the main current forest management techniques used in Brazil, providing a reference, especially for academics, in a simple and summarized way that is easy to understand.

Rio Largo, October 20, 2024.

Eduardo Pagel Floriano

SUMMARY

1 INTRODUCTION

> ***Forest: Land with an area of more than 0.5 hectares with trees more than 5 meters high and a canopy of more than 10 percent, or trees that can reach these thresholds in situ. It does not include land that is predominantly under agricultural or urban use (FAO, 2010).***
>
> ***"Forest management is the scientific ordering of stands for the continuous production of goods and services" (BAKER, 1950).***

Forest management is the science that deals with the sustainable production of tangible and intangible forest goods with the aim of satisfying human needs, based on the principle of optimizing natural resources and their sustainability.

"Forest management" consists of managing forests on a scientific basis to achieve certain objectives, which can be conservation, protection, production, education or recreation (Figura 1).

Humans emerged around 200,000 years ago and since then they have intensively used all the resources of forests in different ways. Forests are not just a community of trees. They also include soil, springs and watercourses, animals and all their flora and fauna. In the beginning, man lived in small groups of gatherers who simply extracted from the environment what it offered naturally. But as civilization evolved, these resources began to be used extensively. In many places, since ancient times, forest resources have become scarce and even extinct. In ancient times, the governments of peoples such as the Chinese and the Romans even issued regulations that disciplined and restricted the use of forests in order to safeguard and maintain forest production in a more sustainable way.

Figura 1 - Main objectives of forests.

Efforts have been made in the past to manage forests and one of the oldest compendiums dealing with the subject is that of Hans Carl von Carlowitz who, in 1713, published the work *Sylvicultura Oeconomica*. He wrote that forests should be used on the basis of their natural characteristics for the well-being of the population, managed and conserved with care and with the responsibility of leaving a good legacy for future generations (GROBER, 2002).

The main focus of forest management is timber production, but other forest services benefit civilization and provide well-being for people, such as shelter and food for fauna, protection of water sources, climate mitigation, carbon sequestration in the air, cloud production, conservation of genetic diversity, food and also leisure and tourism.

Thus, the current concept of forest management can be expressed as the science that deals with the sustainable production of forest goods and services with the aim of satisfying human needs, based on the principle of optimizing natural resources, their conservation and sustainability.

Some historical facts have characterized the evolution of concern for forests and are cited below:

- In 1122 BC, a Chinese Emperor hired a forester to thin, prune and clear stands;
- 23-79 AD, the Romans began to plan the use of forests and were already familiar with the system of management in the high forest and coppice;
- 742 - 814 AD - The beginning of forest management in Central Europe (France, Germany, Austria and Switzerland);
- 1669 - France - the law required that there be an official authorization for all types of felling, and that in the case of final felling, the form of wood cubing, extent and treatment procedure be specified. At this time, sporadic efforts were also made in England;
- During the Middle Ages, wood was an important resource due to its widespread use in domestic construction, shipbuilding and as fuel;

- In Germany in the 14th century, sustainable yield practices were carried out using the area division method. In the 18th century, due to the serious danger of timber shortages, a theory of management was developed. Logging began to be regulated based on the volume rather than the area of the forest;
- In 1713 Hans Calowitz developed the first concept of forest sustainability in the terms that are still understood today;
- The first Forestry School was founded by Hans Dietrich von Zanthier in Wernigerode, Germany, which was closed when its founder died in 1778;
- The Prusian School, founded in 1779 in Hesse by Georg Ludwig Hartig, played a major role in the development of forestry science;
- In 1804, Hartig and Cotta formulated the basic idea of sustainable forest management, which meant managing forests in such a way that descendants would obtain at least the same benefits from them as the current generation;
- In the 19th century, the famous Normal Forest Model was formulated by Hundeshagen and Meyer. This model served as the basis for the majority of cutting regulation methods. Also in this century, many production studies were carried out and various volume and production tables set up, as well as the compound interest calculation carried out by Pressler in 1860;
- The first part of the 20th century was marked by a stagnation in forestry development, caused mainly by the futile struggle between the Net Land Income School, which observes interest on the value of the stand as a cost, and the Net Forest Income School, which does not include interest on the value of stands in the calculation of costs;
- A very promising phase in planning began after the Second World War with the development of operational research, mainly in England and the USA. The mathematical models formulated by this

discipline are, in particular, linear optimization, dynamic optimization, the PERT/CPM system and simulation techniques which, when applied to forest management, allow for more realistic solutions to more complex problems than classic planning techniques.

In Brazil, it can be said that there was no concern with forest management until the middle of the 20th century, although in 1934 the first forestry code was enacted, Decree 23.793/34, which provided for some actions to protect fauna and flora, such as limiting deforestation to a maximum of 75% of the surface area of properties, as well as the replacement of homogeneous forests that had been intensively exploited, and the obligation for steel mills to maintain a sufficient area cultivated with forests to guarantee industrial supply; there was no concern with forest management to improve productivity or guarantee constant production. In 1960, the country had only 500,000 hectares planted, with insipid forest management. But at the beginning of the 1960s, the indiscriminate exploitation of pine forests in the south of the country began to show signs of the need for greater regulation in the sector and improved forest management. In 1965, Law No. 4.771 (the New Forest Code) was enacted, which made management plans for the Amazon forest compulsory, but it wasn't until 1986 that the term sustainable management appeared in Brazilian legislation (Law No. 7.511 of 1986).

On May 30, 1960, the first National School of Forestry was created in Viçosa by Decree No. 48,247. It was transferred to Curitiba in 1963 and graduated Brazil's first class of Forest Engineers on December 8, 1964. With this, Brazilian forestry science began its journey towards the excellence it has achieved in the 21st century.

The cultivation of Brazilian forests for timber production began with Navarro de Andrade, who introduced the *Eucalyptus* in 1903 for the production of railway sleepers (SNIF, 2018).

In 1966, Brazil began a program of tax incentives to create the basis of a forestry production chain with the enactment of Law 5.106, allowing part of the income tax payable by individuals to be invested in forestry, thus beginning the process of forming the forestry base that exists today. Then, in 1970, Decree Law 1134 was passed, which also allowed people to take equity stakes in forestry companies with tax incentives. As a result, around 5.5 million hectares were effectively planted throughout the country by the end of the 1980s. In addition, from the mid-1970s onwards, forestry-based industries began to be set up with government support and, from 2000 onwards, the wood production chain began to walk on its own two feet. According to the IBGE (2018), the areas dedicated to Brazilian forestry totaled around 10 million hectares in December 2016, of which 7.5 million hectares were *Eucalyptus*, 2.1 million hectares *Pinus* and the rest other species (Tabela 1).

The majority of Brazil's planted forests are owned by the pulp industry (34%), followed by independent producers (29%) and then steel mills (14%) (Figura 2).

Recently, monoculture forestry has given way to new plantations in agroforestry and silvopastoral systems with positive economic results that are less aggressive to the environment and more socially just. In Figura 3 shows the distribution of planted forests by state.

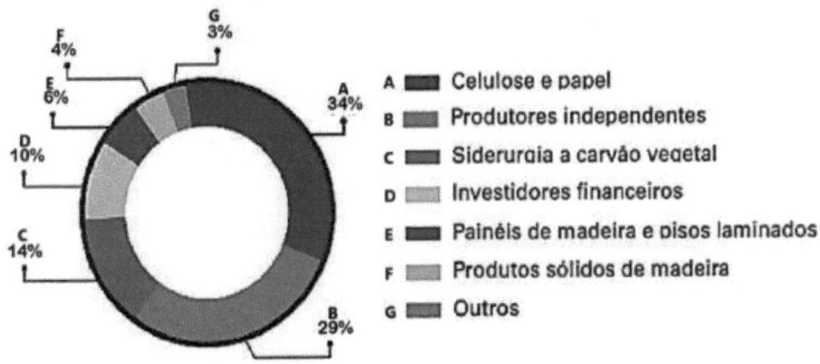

Figura 2 - Forest plantations by type of owner. Source: IBA and POYRY (2017).

Planted forests in Brazil have many desirable aspects, such as reducing pressure on native forests, restoring land degraded by livestock and agriculture, greater carbon sequestration than most rural activities, protection of water sources and soil, shorter rotation cycles compared to countries with a temperate climate and lower risk of climate variations compared to annual crops.

Brazilian forestry is distributed in the Southeast (34%), South (37%), Northeast (9%), North (6%) and Midwest (14%), as shown in Tabela 1 (IBGE, 2018).

The initial productivity of *Eucalyptus* plantations, back in the 1970s, was no more than 25 $m^3.ha^{-1}.year^{-1}$ and today the average is over 35 $m^3.ha^{-1}.year^{-1}$, while some companies are already reaching 60 $m^3.ha^{-1}$ $.year^{-1}$ in 7-year *Eucalyptus* rotations (SNIF, 2018).

Tabela 1 - Forestry area by federation unit, by species .

Federation Units, Mesoregions, Microregions and Municipalities	Existing area of forestry stocks on 31.12.2016 (ha)			
	Total	Forest species		
		Eucalyptus	Pine	Other species
North	602,403	511,772	5,291	85,340
Rondônia	28,000	2,600	4,800	20,600
Acre	-	-	-	-
Amazonas	-	-	-	-
Roraima	10,390	-	-	10,390
Pará	201,714	154,907	-	46,807
Amapá	221,252	219,545	48	1,659
Tocantins	141,047	134,720	443	5,884
North East	908,832	900,757	575	7,500
Maranhão	261,616	261,605	-	11
Piauí	36,316	36,316	-	-
Ceará	270	7	-	263
Rio Grande do Norte	461	-	-	461
Paraíba	6,084	1,040	-	5,044
Pernambuco	1,291	228	-	1,063
Alagoas	11,967	11,337	-	630
Sergipe	3,363	3,335	-	28
Bahia	587,464	586,889	575	-
South East	3,363,590	3,129,918	227,207	6,465
Minas Gerais	1,880,538	1,839,459	38,933	2,146
Holy Spirit	289,376	287,057	2,047	272
Rio de Janeiro	37,373	36,552	8	813
São Paulo	1,156,303	966,850	186,219	3,234
South	3,736,702	1,678,478	1,832,974	225,250
Paraná	1,635,583	684,382	920,251	30,950
Santa Catarina	1,015,801	341,130	647,322	27,349
Rio Grande do Sul	1,085,318	652,966	265,401	166,951
Midwest	1,411,549	1,322,782	13,115	75,652
Mato Grosso do Sul	998,083	993,807	4,276	-
Mato Grosso	266,017	191,995	-	74,022

Federation Units, Mesoregions, Microregions and Municipalities	Existing area of forestry stocks on 31.12.2016 (ha)			
	Total	Forest species		
		Eucalyptus	Pine	Other species
Goiás	144,049	134,280	8,139	1,630
Federal District	3,400	2,700	700	-
Brazil	10,023,076	7,543,707	2,079,162	400,207

Source: IBGE (2018).

Other species besides *Eucalyptus* and *Pinus* have also been cultivated for specific purposes in different regions of the country, such as black wattle (*Acacia mearnsii*), rubber (*Hevea spp.*), teak (*Tectona grandis*), paricá (*Schizolobium parahyba*), *araucaria* (*Araucaria angustifolia*), poplar (*Populus sp.*), Australian cedar (*Toona ciliata*) and African mahogany (*Khaya ivorensis* and *Khaya anthotheca*).

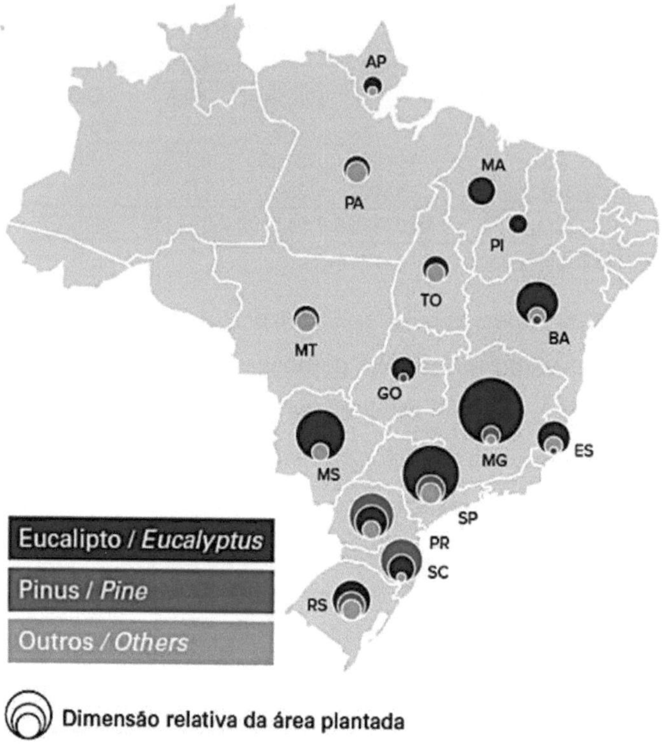

Figura 3 - Forest plantations by state. Source: IBA (2018).

According to IBGE (2023), the main products obtained from natural forests (extraction) and planted forests (forestry) in Brazil are charcoal, firewood and logs (Figura 4). Consequently, different forms of management are applied to each type of forest, in each location, in order to meet production and market supply objectives. Considering a 35% yield in the carbonization of wood with a density of 0.5, the production of these products reached around 300 million cubic meters, with forestry accounting for more than 89% of total production.

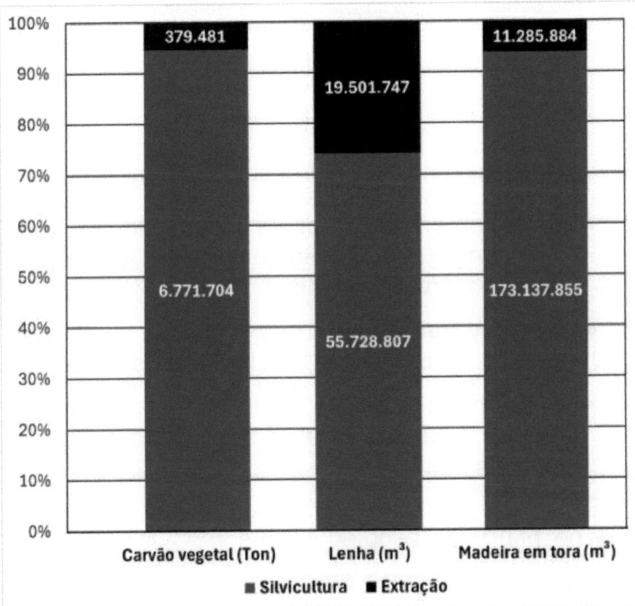

Figura 4 - Production by type of product and forest in Brazil in 2023. Source: IBGE (2023).

1.1 Requirements for sustainable forest management

Requirements are the conditions required for forest management to be sustainable, optimizing resources and maximizing production with minimum negative environmental impact and maximum positive socio-economic impact. A lot has been written about these requirements in a disorganized way in the past. Forest certification systems have organized them into clear standards. The two best-known certification systems in Brazil are the Brazilian Forest Certification

Program (Cerflor) and the *Forest Stewardship Council* (FSC) .[1]

Forest certification programs adopt three basic requirements, demanding that forest management be:

- **Environmentally Appropriate** - Protecting and maintaining natural communities and forests of high conservation value;
- **Socially Beneficial** - Respecting the rights of workers, communities and indigenous peoples;
- **Economically Viable** - Building markets, adding greater value and creating equitable access to benefits.

The principles adopted by the FSC for good forest management are:

- Principle 1 - Compliance with the Laws;
- Principle 2 - Workers' Rights and Working Conditions;
- Principle 3 - Rights of Indigenous Peoples;
- Principle 4 - Community Relations;
- Principle 5 - Benefits of the Forest;
- Principle 6 - Environmental Values and Impacts;
- Principle 7 - Management Planning;
- Principle 8 - Monitoring and Evaluation;
- Principle 9 - High Conservation Values;
- Principle 10 - Implementation of Management Activities.

Cerflor adopts five basic requirements in the form of principles, as follows:

---

[1] Forest Management Council

"Principle 1 - Compliance with legislation, the forest enterprise must be managed through attitudes and actions that ensure compliance with federal, state and municipal legislation, as well as international agreements, treaties and conventions applicable to forest management. The organization must provide adequate protection of the forest, in order to prevent unauthorized activities, such as illegal logging and other illegal activities, and to respect the legislation applicable to forest management issues, such as environmental protection, endangered and protected species, land tenure, ownership and use rights for indigenous peoples and traditional communities, labour and health and safety issues, and the payment of royalties and taxes.

Principle 2 - Rational use of forest resources in the short, medium and long term, in pursuit of sustainability, forest management planning should aim to maintain the health and vitality of forest ecosystems, seeking to maintain and increase the economic, ecological, cultural and social values of the forest. The forest should be managed in such a way that the activity contributes to the conservation of renewable natural resources.

As Principle 3 - Care for biological diversity, the organization must manage the forest plantation in such a way as to minimize the negative impacts of its forestry activity on native flora and fauna. It must ensure that biological diversity is maintained and increased.

Principle 4 - Respect for water, soil and air the forest management and technological development program indicates to companies that they should consider and adopt techniques that take into account soil, water and air conservation.

Principle 5 - Environmental, economic and social development of the regions in which forestry activities are carried out, requires the company to have a policy of relations with forestry workers and local communities, as well as evidence of the benefits of forestry activities in social, environmental and economic terms (TARGET, 2013). "

Cerflor is internationally accredited by the *Program for the Endorsement of Forest Certification Schemes* (PEFC)[2] , the body that carries out the international accreditation of forest certification systems.

According to PEFC, sustainable forest management is a holistic approach defined as the management and use of forests and forest areas in a manner and at a rate that maintains their biodiversity, productivity, regeneration capacity, vitality and potential to fulfill, now and in the future, relevant ecological factors, economic and social functions, at local, national and global levels, without causing damage to other ecosystems.

1.1.1 General requirements for certification

Regulatory requirements for sustainable forest management defined at regional, national or sub-national level must:

- Include management and performance requirements that are applicable at the forest management unit level, or other appropriate level, to ensure that the objectives of all requirements are achieved at the forest management unit level; note: An example of a situation where a requirement can be defined as being other than at the forest management unit level (e.g. group / region) is forest health monitoring. By monitoring forest health at the regional level and reporting the results at the FMU level, the objective of the requirement is met without the need to carry out individual monitoring of each forest management unit;

---

2 Program for endorsement of forest certification systems (http://www.pefc.org).

- Be clear, objective and auditable;
- Apply to the activities of all operators in the defined forest area that have a measurable impact on compliance with the requirements;
- Require the keeping of records that provide evidence of compliance with the requirements of the forest management standards.

1.1.2 Specific criteria for certification:

- Criterion 1: Maintenance and appropriate enhancement of forest resources and their contribution to the global carbon cycle;
- Criterion 2: Maintaining the health and vitality of the forest ecosystem;
- Criterion 3: Maintaining and encouraging the productive functions of forests (timber and non-timber);
- Criterion 4: Maintenance, conservation and appropriate improvement of biological diversity in forest ecosystems;
- Criterion 5: Maintenance and appropriate enhancement of protective functions in forest management (mainly soil and water);
- Criterion 6: Maintenance of other functions and socio-economic conditions;
- Criterion 7: Compliance with legal requirements.

1.2 Elements of forest management

The six main elements to consider in forest management are listed below:

- Spatial - productive, conservation and preservation spaces;
- Temporal - cycles, rotation;
- Economic - market, resources and production objectives;

- Ecological - environment (soil, air, water, climate, topography) for production, conservation and preservation;
- Political - laws, regulations, public policies;
- Socio-cultural - labor, technology.

2 APPROACHES TO FOREST MANAGEMENT

The three different objectives of forests (economic, social and environmental) lead to different approaches and forms of management.

2.1 Forests for conservation purposes

In Brazil they are called Conservation Units and their management is regulated by Law No. 9.985 of July 18, 2000, which divides them into two groups: full protection units and sustainable use units.

2.1.1 Full Protection Units

The aim of Full Protection Units is to preserve nature, with only indirect use of its natural resources permitted.

The following categories make up the group of Full Protection Units:

- I - Ecological Station;
- II - Biological Reserve;
- III - National Park;
- IV - Natural Monument;
- V - Wildlife Refuge.

2.1.2 Sustainable Use Units.

The aim of Sustainable Use Units is to make nature conservation compatible with the sustainable use of part of its natural resources.

IX - on hilltops, hills, mountains and mountain ranges, with a minimum height of 100 (one hundred) meters and an average slope greater than 25°, the areas delimited from the contour line corresponding to 2/3 (two thirds) of the minimum height of the elevation always in relation to the base, this being defined by the horizontal plane determined by the adjacent plain or water mirror or, in undulating reliefs, by the elevation of the saddle point closest to the elevation;

X - areas at an altitude of more than 1,800 (one thousand eight hundred) meters, whatever the vegetation;

XI - in veredas, the marginal strip, in horizontal projection, with a minimum width of 50 (fifty) meters, starting from the permanently marshy and waterlogged space.

§ Paragraph 1 Permanent Preservation Areas will not be required around artificial water reservoirs that are not the result of damming or impounding natural watercourses.

§ Paragraph 2 (Repealed). (Redaction given by Law no. 12.727, of 2012).

§ PARAGRAPH 3 (VETOED).

§ Paragraph 4. In natural or artificial accumulations of water with a surface area of less than one (1) hectare, the reservation of the protection strip provided for in items II and III of the caput is dispensed with, and new suppression of areas of native vegetation is prohibited, unless authorized by the competent environmental agency of the National Environmental System - Sisnama. (Editing given by Law No. 12.727, of 2012).

> *§ Paragraph 5 - The planting of short-cycle temporary and seasonal ebb and flow crops on the strip of land that is exposed during the ebb and flow period of rivers or lakes is permitted for smallholdings or rural family possessions, as referred to in Article 3(V) of this Law, provided that it does not involve the suppression of new areas of native vegetation, that the quality of the water and soil is conserved and that wild fauna is protected.*
>
> *§ Paragraph 6 - In rural properties with up to 15 (fifteen) fiscal modules, the practice of aquaculture and the physical infrastructure directly associated with it is permitted in the areas referred to in items I and II of the main body of this article, provided that:*
>
> *I - sustainable soil and water management practices and water resources are adopted, guaranteeing their quality and quantity, in accordance with the rules of the State Environmental Councils;*
>
> *II - is in accordance with the respective basin plans or water resource management plans;*
>
> *III - licensing by the competent environmental agency;*
>
> *IV - the property is registered in the Rural Environmental Registry (CAR)."*

Legal Reserve areas and those used for landscaping or climate mitigation are also considered protection forests. The Legal Reserve (RL) is defined in Law 12.651/2012 as: an area located within a rural property or possession, delimited under the terms of art. 12, with the function of ensuring the sustainable economic use of the rural property's natural resources, assisting the conservation and rehabilitation of ecological processes and promoting the conservation of biodiversity, as well as the shelter and protection of wild fauna and native flora.

Some protective forests can be productive, such as legal reserves, but they must be managed using a low-impact system and without clear-cutting. The others are managed solely to preserve them and, when they are degraded, management must also aim to restore them.

Any activity in protected forests, defined by law as APP or RL, requires prior authorization from the competent official body.

2.3 Forests for production purposes

Productive forests are all those that are used to produce goods with monetary value. The management of these forests has been carried out with different objectives, which imply different sizes of wood to be obtained and different scales of production. Depending on the scale of production and the objectives of the wood, forests have been managed at different intensities, from extensive management without technical-scientific guidance to intensive management to maximize productivity or profitability (Tabela 2).

Tabela 2 - Intensity of forest management in Brazil according to timber objectives and scale of production

Purpose of the wood	Scale of production			
	Small	Average	Large	
	Market			
	Location	Regional	National/World	
General use on the farm	Low	-	-	-
Domestic energy and small industry	Low	Average	-	-
Rough sawn timber for buildings	Low	Average	Intensive	Intensive
Sawmills and veneers	Average	Average	Intensive	Intensive
Coal for the steel industry	-	Average	Intensive	Intensive
Reconstituted wood panels	-	-	Intensive	Intensive
Cellulose	-	-	Intensive	Intensive

The dimensions and uses of the wood imply appropriate silvicultural techniques that will be part of the management of the stands, such as the definition of initial spacing, pruning and thinning.

Wider spacings are used to obtain large wood for sawmills and veneers, while the wood to be used in fiber, pulp and energy production is produced with smaller spacings. Smaller spacings allow the canopy to close earlier and make the trees grow less individually and have a higher productivity per unit area, which is an advantage in shorter rotations. The trunks formed in widely spaced stands are less cylindrical or more conical and result in a greater amount of waste when sawing or turning the wood. However, the individual growth of the trees in diameter is greater, partially compensating for the losses in the industrialization of the wood, or even with a surplus of sawn timber compared to trees that grew less widely spaced and with trunks that are more cylindrical in shape but smaller in size.

Pruning is carried out to produce solid wood, which is knot-free and has a better appearance when exposed to view, but it is generally unnecessary in the production of wood for energy, pulp or fiber.

Small wood is produced in short rotations without thinning, while medium and large wood is produced in longer rotations with thinning. Wider spacings and little thinning shorten the rotation for the production of thick wood, but the branches develop more and in greater numbers, which leads to the implementation of slightly less wide initial spacings associated with early thinning.

3 PHASES OF FOREST MANAGEMENT

Forest management should include a number of phases, such as the ones below, which can serve as guidelines for those who want to produce wood sustainably. The first phase is to gather information, then there is the definition of objectives, characterization of the enterprise, definition of species and dimensions of wood, growth studies and production planning, economic and feasibility studies, definition of rotation age, prognosis of social, environmental and economic impacts, identification of the risks and difficulties of the enterprise and creation of a monitoring system for the forests, the natural environment, infrastructure and operations to provide sufficient data for verification and correction.

Most decision-making processes in forest management include the following steps (BETTINGER, 2009):

1) Allow the public to participate and comment on the management of an area.
2) Determine the goals for a management area.
3) Inventory the conditions necessary to evaluate the objectives.
4) Analyze trends in land use change and vegetative growth.
5) Formulate alternatives for the area.
6) Evaluate the alternatives for the area.
7) Select an alternative and develop a management plan.
8) Implement the management plan.
9) Monitoring the management plan.

10) Update the management plan.

The steps for drawing up a management plan can be as follows:

1) Study of the socio-economic environment.
2) Study of the biophysical environment.
3) Characterization of the production chain.
4) Market research.
5) Identification of objectives:
   a. production
   b. social
   c. environmental
6) Characterization of the organization (enterprise):
   a. location and scope.
   b. areas: productive, preserved, infrastructure.
   c. shareholders and participation quotas.
   d. area of influence.
7) Choice of species and dimensions.
8) Production planning:
   a. Study of growth.
   b. Study of site quality.
   c. Study of competition.
   d. Studying the density of stands.
   e. Modeling forestry production.
   f. Determining production costs.
   g. Determining the revenue to be obtained.
9) Economic feasibility study.
10) Definition of the rotation age.
11) Prognosis of economic, environmental and social impacts.
12) Identifying risks and difficulties.
13) Expected results.
14) Forestry and operational control and monitoring system.

3.1 Socio-economic environment

3.1.1 Identifying the actors

The actors in a forestry production plan are represented by the leaders of organized society, including local, regional and national society, development agencies, state secretariats for planning, management, infrastructure, agriculture, forestry and agrarian development, environmental agencies, teaching, research and extension institutions, environmental and assistance institutions, councils, financial agents, research and development funds, among others.

3.1.2 Socio-economic characterization

This phase involves studying regional and other levels of economic growth, the evolution of socio-economic indicators, the characteristics of the population, the urban network, sanitation indices, land availability and prices, regional agriculture, planted and native forests, imports and exports, the existing forest products market and its forecasts. It is also important to collect data on education at all levels, identify the availability of labor, employment and unemployment rates,

3.1.3 Development policies

It is important to identify government development policies that interface with the forestry area and look for ways to fit in with them as a way of obtaining exemptions, loans, cost reductions and seeking participation.

One example is the National Forest Program (PNF) created by Decree No. 3,420 of April 20, 2000, with the aim of articulating sectoral public policies to promote the sustainable development, use and conservation of Brazilian forests, consisting of projects designed and executed in a participatory and integrated manner by the federal, state, district and municipal governments and organized civil society.

3.1.4 Infrastructure

Studying the existing public infrastructure is essential to the success of any development.

The first step in this direction is to identify the modes of transport available to supply forestry, transport products from the forest to industry or the market, and the means of transport available for access by employees, suppliers and customers. All modes of transportation must be identified: cycle paths, roads, waterways, railroads, ports and airports.

This is followed by the study of the production and distribution of electricity to supply production and the availability of water for forestry and industrial production.

3.2 Benchmarking

This is a phase that many forget, which can be fatal to the failure of an enterprise. Visiting, investigating and analyzing other organizations with similar objectives is what allows an organization to situate itself in the market in a constructive and competitive way.

Comparative analysis makes it possible to identify the best references and practices carried out by organizations in the same market range in all areas: who their investors are, who their employees are and what their people management is like, who their suppliers and customers are and how they relate to them, what the quality of their products and prices are, what production equipment they use, what their productivity and costs are, etc.

3.3 Biophysical environment

Characterizing the biophysical environment allows for the correct choice of genetic strains, silvicultural practices and forest management. It also helps determine which sites are suitable for forestry and which should be prioritized for agricultural production, as well as which areas should be set aside for preservation and conservation.

When characterizing the biophysical environment, the physiography and natural vegetation must be characterized, land use and cover identified, private and public conservation units, government environmental zoning, land availability and occupation trends, soil types and their fertility, hypsometry, land slope, hydrology, climate (temperatures, precipitation, evapotranspiration, insolation, water balance, etc.).

3.4 Characterization of the production chain

The main forest-based production chains are represented by production for:

- Solid wood industries: sawn timber, laminating and splitting;
- Cellulose industries: short fiber (angiosperms) and long fiber (gymnosperms);
- Wood energy production: thermoelectric plants;
- Charcoal production: for steel mills and domestic consumption;
- Boards: MDF (Medium Density Fiberboard), MDP (Medium Density Particleboard), OSB (Oriented Strand Board) and plywood (Figura 5);
- Construction industry (various forest products, mainly boards and sawn timber);
- Furniture industry (various forest products), mainly boards;
- Chemical industries: resins, paints, disinfectants, flavorings, medicines, cosmetics.

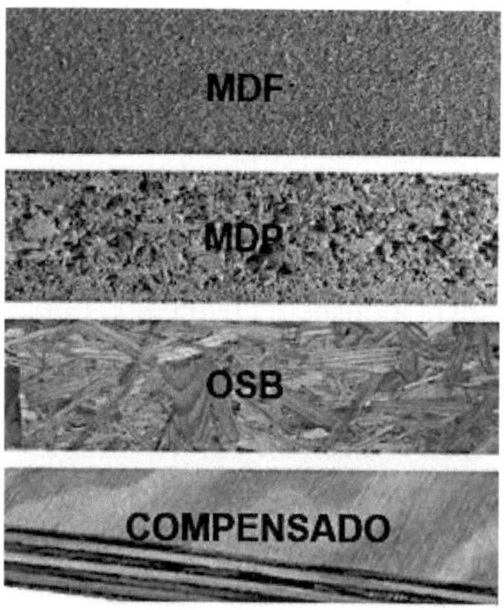

Figura 5 - Fibreboard (MDF), particleboard (MDP), oriented strand board (OSB) and plywood (cross-ply).

3.4.1 Characteristics of forest-based industries

The figures listed below are approximate and may vary according to the type of enterprise, technology, location, investment, etc.

This topic discusses the main aspects of the different sub-sectors of the wood production chain in terms of the needs and productivity for implementing forest-based enterprises, such as:

- Minimum economic output; minimum volume of production considered sustainable and economically viable in the long term.
- Water consumption: amount of water needed to produce a unit.
- Energy consumption: the amount of energy needed to produce a unit.
- Chemical consumption: the quantity of bulk chemical products needed to produce a unit.
- Wood consumption: volume of wood needed to produce a unit.
- Cultivation area: effective planting area required for minimum economic production.
- Total area: total area required including production areas, reserves and permanent preservation areas to achieve minimum economic production.
- Labor: the number of workers required to achieve minimum economic output.
- Investment: average investment value per production unit.

3.4.2 Productivity Industrial

The productivity of Eucalyptus wood can vary greatly in terms of pulp production. Investment in the development of suitable clones has considerably improved the yield of planted forests, which currently exceeds 10 tons of pulp per hectare per year in most cases. In Tabela 3 two scenarios are presented, the first referring to the yields that were obtained in the 1970s/80s and the second referring to a current optimum situation, which the best companies are achieving.

Tabela 3 - Average yields of seven-year-old *Eucalyptus* plantations for pulp production.

Features	Scenario 1 - worst	Scenario 2 - better
Productivity in volume (m3/ha/year)	30	45
Cellulose yield (%)	48	50
Consumption at the mill (m3/t of pulp)	4,5	3,5
Forest productivity (t cel./ha/year)	6,67	12,86

The number of jobs generated by the planted forest sector in Brazil in 2016 was around 510,000, with an estimated 3.7 million indirect jobs.

The variation in labor productivity, in terms of value produced per worker per year, shows the phase of technological change that the country is going through, where modern industries with advanced technology coexist with other artisanal and unproductive ones that gradually tend to disappear, as can be seen in Tabela 4 .

Tabela 4 - Labor productivity in forest-based subsectors .

Activity	Labor productivity (R$ / worker)		
	Average	Lower limit (10% less competitive)	Upper limit (10% most competitive)
Wood splitting	12.246	334	32.749
Slats and panels	21.985	85	54.043
Furniture	11.422	192	11.422

Source: IBGE (2005).

Tabela 5 - Evolution of jobs generated by forestry segment in Brazil

Forestry sector	2006	2007	2008	2009	2010	2011	2012	2013	2014	2015
Support activities for forestry production [1]	59197	60787	52376	44419	53069	54504	47289	39909	38530	33049
Splitting wood [2]	100982	99183	87929	83114	87586	85215	81267	78078	75734	70654
Pulp and paper production	154419	158676	161354	163182	173219	175122	177230	181634	184767	177323
Production of wooden structures and artifacts	44386	45407	45061	43742	47559	48481	48688	48402	47540	44013
Production of wood veneers and sheets	52200	50786	45086	39491	42045	41208	40644	40888	40563	37869
Forestry production - Forests native [3]	8744	8671	6443	6382	7160	8189	8380	7380	8295	7668
Forestry production - Planted forests [4]	51406	62499	65454	62877	69474	70316	66734	64543	62519	63058
Furniture production [5]	160117	168139	171218	172740	188178	196647	204743	207208	208481	191929
Total	631451	654148	634924	615947	668290	679682	674975	668042	666429	625563

Where: * Number of active jobs on 12/31. 1 Support activities for forestry production: includes logging, felling and transportation services, wood assessment, dendrometry, wood unloading and services related to forestry and plant exploitation. 2 Wood splitting: includes the production of floors, flooring, sleepers, planks, linings, boards, wooden poles, etc. and the processing of sawn timber. 3 Forestry production - native forests: includes extraction, felling, collection, charcoal production and processing activities. 4 Forestry production - planted forests: includes cultivation activities, production of seedlings, bark, leaves and resins, extraction, felling and reforestation with felling of trees. 5 Furniture production: includes the manufacture of furniture with a predominance of wood. Source: SNIF (2018).

A more consistent and comprehensive study than that of the Tabela 4 was carried out by Najberg and Pereira (2004) and is shown in Tabela 6 .

Tabela 6 - Jobs Generated by Increased Production of R$10 million, at average 2003 prices (Najberg and Pereira, 2004).

Setor	Number of jobs							
	Direct	Rank	Indirect	Rank	Income effect	Rank	Total	Rank
Agriculture	393	4	131	15	303	7	828	3
Mineral extract Mineral	90	14	126	17	266	20	481	21
Oil and gas	9	38	84	30	329	2	422	25
Mineral ñ metallic	99	12	117	20	261	21	477	22
Steelmaking	8	39	135	14	259	22	402	27
Metallurgy Metallurgy	18	34	97	28	202	40	316	38
Other metalworkers	98	13	109	22	244	27	451	23
Machinery and equipment	62	17	80	34	278	14	420	26
Electrical equipment	37	25	121	18	213	34	371	31
Equip. Electronics	41	22	83	32	208	36	332	35
Cars/cams/buses	16	35	108	24	203	39	326	37
Parts and out. Vehicles	37	26	117	21	234	30	387	28
Wood and furniture	**293**	**6**	**219**	**8**	**294**	**8**	**805**	**5**
Pulp, paper and graphics	**59**	**19**	**155**	**11**	**271**	**17**	**485**	**20**

Setor	Number of jobs							
	Direct	Rank	Indirect	Rank	Income effect	Rank	Total	Rank
Rubber industry	23	32	108	23	229	31	360	33
Chemical elements	14	37	188	9	289	11	491	19
Oil refining	2	41	62	38	208	37	271	41
Various chemicals	26	31	99	26	213	35	339	34
Pharmac. And veterinary	38	24	117	19	222	33	377	30
Plastic articles	88	15	68	36	206	38	362	32
Textile industry	62	18	144	12	176	41	382	29
Articles of clothing	613	2	136	13	250	25	1000	2
Footwear manufacturing	246	7	174	10	290	9	711	7
Coffee industry	41	23	356	2	323	3	719	6
Benef. Prod. Vegetables	58	20	327	4	259	23	643	11
Slaughtering animals	36	27	358	1	270	18	664	9
Dairy industry	29	30	326	5	267	19	621	13
Sugar production	32	29	307	6	337	1	677	8
Fab. Vegetable oils	8	40	350	3	284	13	642	12
Other food products Food	82	16	238	7	252	24	572	14
Various industries	124	11	126	16	250	26	501	18
S.i.u.p.	21	33	41	40	238	28	299	40
Construction	176	9	83	33	271	16	530	17
Trade	449	3	84	31	278	15	810	4
Transport	219	8	96	29	237	29	551	16

Setor	Number of jobs							
	Direct	Rank	Indirect	Rank	Income effect	Rank	Total	Rank
Communications	33	28	45	39	227	32	305	39
Financial institutions	47	21	80	35	310	5	437	24
Serv. Services Family	665	1	104	25	311	4	1080	1
Serv. Company	293	5	63	37	288	12	645	10
Renting property	15	36	10	41	307	6	331	36
Public administration	165	10	98	27	290	10	553	15

Source: Employment generation model - BNDES. Data sources: CN02, MIP96, PNAD01, POF95/96. Last updated: February 2004.

EUCALYPTUS PULP, KRAFT PROCESS

1) Minimum economic production: 700,000 tons per year;
2) Average price of 1 ton of pulp: US$800.00;
3) Water consumption: 25 m³ of water per ton of pulp;
4) Energy consumption: 0.784 mW per ton of pulp;
5) Chemical consumption: Chlorine, 25 kg/t; Soda, 25 kg/t;
6) Wood consumption: 3.7 m³ of wood per ton of pulp;
7) Cultivation area (35 m³/ha/year): 74,000 hectares;
8) Total area (with native reserves): 110,000 hectares;
9) Labor: forestry, 2 workers / 100 ha; industrial, 1 worker / 1 thousand tons per year;
10) Investment: US$ 1,800 / ton of pulp per year.

SAWN TIMBER

1) Minimum economic output: 100,000 m³ per year;
2) Average price of sawn timber: US$160/m³;
3) Water consumption: 0.1 m³ of water per m³ of wood;
4) Energy consumption: 2 mWh / m³ of sawn timber;
5) Wood consumption: 2.42 m³ of standing timber / m³ sawn;

6) Cultivation area (30 to 32 m³/ha/year): 8.1 thousand hectares;
7) Total area (with native reserves): 13,500 hectares;
8) Labor: forestry, 3 workers / 100 ha; industrial, 5 workers / 1000 m³ per year;
9) Investment: US$ 300 / m³ sawn per year.

LAMINATION

1) Minimum economic output: 50,000 m³ per year;
2) Water consumption: 5 m³ of water per m³ of laminated wood;
3) Energy consumption: 265 kW / m³ of dry laminated wood;
4) Wood consumption: 3 m³ of standing timber / m³ of laminate;
5) Cultivation area (30 to 32 m³/ha/year): 5,000 hectares;
6) Total area (with native reserves): 8.3 thousand hectares;
7) Labor: forestry, 3 workers / 100 ha; industrial, 7 workers / 1000 m³ per year;
8) Investment: US$ 330 / m³ of annual slates.

According to Revista Expressiva (2005), an industry that will produce 6,000 m³ of laminated wood per month in Santa Catarina has announced a total investment of R$10 million and the creation of 500 direct jobs.

WOOD FIBER PANELS

1) Minimum economic output: 250,000 m³ per year;
2) Average price of fiberboard: US$400.00/m³;
3) Water consumption: 20 m³ of water per m³ of boards;
4) Energy consumption: 0.5 mWh / m³ of boards;
5) Chemical consumption: 8% tannin-formaldehyde

resin;

6) Wood consumption: 1.8 m^3 of standing timber / m^3 of panels;
7) Classification: low density <0.6 g/cm^3; medium density >=0.60 and <0.8 g/cm^3; high density >0.80 g/cm^3;
8) Cultivation area (35 m^3/ha/year): 13,000 hectares;
9) Total area (with native reserves): 20,000 hectares;
10) Labor: forestry, 3 workers / 100 ha; industrial, 2 workers / 1000 m^3 per year;
11) Investment: US$ 600 / m^3 of panels per year.

THERMOELECTRIC

1) Minimum economic output: 5 MWh;
2) Value of energy generated: US$95/MWh;
3) Water consumption: 10 m^3 per MWh;
4) Energy consumption: self-generated;
5) *Eucalyptus* calorific value: 1915 kWh / m^3;
6) Wood consumption: (13% generation losses) 0.6 m^3 per Wh;
7) Cultivation area (35 m^3/ha/year): 2.4 thousand hectares;
8) Total area (with native reserves): 3.6 thousand hectares;
9) Labor: forestry, 3 workers / 100 ha; industrial, 3 workers / MWh produced;
10) Investment: US$ 1.73 million / MW per year.

According to Lopes and Wipieski (2001), a module with an effective area of 2,400 ha of planted *Eucalyptus* forest has the capacity to sustainably produce around 5 MW/year and would generate around 200 jobs in the forestry, harvesting and operations areas, representing an overall investment of approximately R$8 million. They also report that a sawmill with an annual production of around 36,000 m^3 of sawn timber could generate between 10,800 and 13,200 MWh/year from waste, enough to meet its own energy demand.

In 2014, the website "Celulose Online" published the announcement that the largest thermoelectric plant in Latin America would be set up, with an investment of R$650 million (around US$260 million at the time), to generate 150MW, requiring a 230 kV transmission line over a distance of 90 km to integrate it into the national system (SIN), at an estimated cost of around US$23 million. For the production of *Eucalyptus* biomass, around 35,000 hectares of plantations have been planned. The plant is also expected to generate 800 direct and indirect jobs.

3.5 Choice of species

The main factors determining the choice of species are the production objectives, the adaptation of the species and provenances to the regional environment and production objectives, as well as productivity and profitability.

3.5.1 Production objectives

The production objectives must be established according to the market for the product you want to produce and the choice of species must take into account its suitability for the products you want to obtain. The IPT (Institute for Technological Research) maintains information on wood and its applications, which can be consulted online at https://www.ipt.br/consultas_online.

The genera of trees most commonly used in commercial plantations are *Pinus* and *Eucalyptus*. *Eucalyptus* species are suitable for a wide variety of products, including:

- Construction: beams, rafters, rulers, props, scaffolding and roof structures;
- Energy: firewood, charcoal;
- Short-fiber cellulose: printing and hygiene papers;
- Soluble cellulose: used in the production of hydroxyethyl cellulose, carboxymethyl cellulose, cellulose gum, micro crystalline cellulose, regenerated cellulose, used in various segments listed below. Textiles: manufacture of viscose. Automotive industry: high resistance tires. Chemical industry: paints, varnishes, explosives. Food: it is used in candy wrappers, chewing gum, sausage casings; in food composition to give creaminess to soft drinks, diet shakes, ice cream, mayonnaise and ketchup, sauces, soups and sweets. Other: cigarette filters, eyeglass frames, contact lenses, photographic film, toothpaste, soluble medicine capsules, soaps, LCD (liquid crystal display) screens, etc;
- Essential oil: it is used in the cosmetics and cleaning products industry, in the composition of medicines as a tranquillizer, relaxant, sleep inducer, it has antibacterial and antifungal properties in the healing of wounds, burns or insect bites; it is used in the composition of products for oily skin, decreasing the secretion of the sebaceous glands.

The *Pinus* species grown in Brazil are mainly *P. elliottii*, *P. taeda* and *P. caribaea*. All three species are used for timber production and only *P. taeda* is not used for resin production.

Pinus wood is soft and easy to work with. Its main applications are as follows. Construction: internal exposed beams; laths; secondary parts of structures; stringers; trims; skirting boards; ceilings; paneling; internal doors; concrete forms; pontoons. Furniture: solid wood furniture; internal parts of furniture. Other uses: broom handles; plywood sheets; decorative veneers; turned parts; sporting goods and toys; packaging; bobbins and spools; brushes.

The main by-products of *pine* resin are rosin and turpentine. Rosin, the main by-product of gum resin, is obtained from three sources, as shown in Figura 6 The main by-products of gum-resin are stump rosin (extracted with solvents from residues and stumps left over from harvesting the wood), *tall oil* rosin (obtained from black lye, which is a residue from the manufacture of cellulose) and rosin produced from gum-resin extracted from living trees, which has many applications, such as: glues used in the manufacture of paper, paints and varnishes, hot melt, adhesives, synthetic rubbers; the cosmetics and food industries, among others (NEVES et al., 2001).

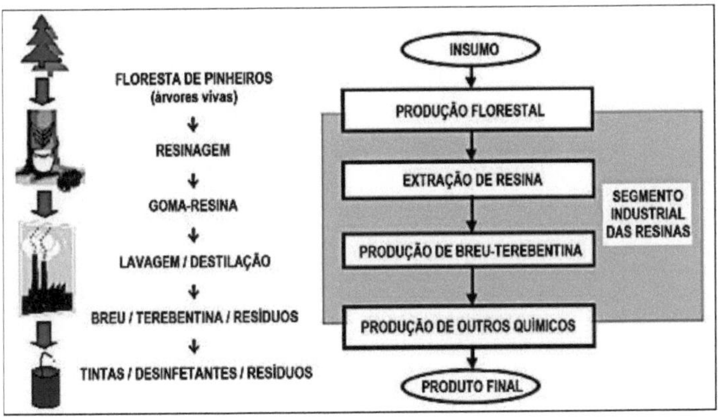

Figura 6 - Agro-industrial system for gum resin and derivatives. Source: Adapted from Ferreira (2001).

Other species are cultivated on a smaller scale, such as *Acacia* mearnsii (black wattle), grown in Rio Grande do Sul to produce tannin and firewood; teak, grown in the Midwest to produce wood for sawing and veneering; Australian cedar (*Toona ciliata*) and African mahogany (*Khaya spp.*), grown in the Southeast, also for sawing and veneering, used as wood for carpentry and joinery.

Among the native species, Carvalho (2003-2014) describes in five volumes, under the title "Brazilian Tree Species", the characteristics and potential of each one. The characteristics of some of the species described by the author are available at <http://www.cnpf.embrapa.br/pesquisa/efb/>, where they are listed as potential or promising, with an average annual increase in volume of more than 10 $m^3.ha^{-1}.year^{-1}$, the 21 species listed in Tabela 7 .

Tabela 7 - Promising native species for timber production.

Scientific name	Popular name	IMA (up to)	Age (years)	Apparent specific mass (g/cm3)
Hymenaea courbaril var. stilbocarpa	Jatobá	10.25	10	0,90 a 1,10
Nectandra lanceolata	Yellow cinnamon	10.4	9	0,47 a 0,70
Balfourodendron riedelianum	Ivory wood	12	26	0,80 a 0,90
Colubrina glandulosa var. reitzii	Sobrasil	12.9	7	0,80 a 1,00
Parapiptadenia rigida	Angico-gurucaia	13.4	12	0,75 a 1,00
Cabralea canjerana subsp. canjerana	Canjarana	13.5	10	0,45 a 0,75
Centrolobium robustum	Araribá-rosa	13.7	9	0,70 a 0,80
Prunus brasiliensis	Varoveira	14.45	10	0,69 a 0,92
Talauma ovata	Baguaçu	15.45	15	0,56 a 0,65
Caesalpinia leiostachya	Ironwood	17.2	14	0,99 a 1,27
Cariniana estrellensis	White Jequitibá	17.2	25	0,70 a 0,78
Myracrodruon balansae	Southern ironwood	18	28	1,10 a 1,25
Centrolobium tomentosum	Araruva	19.3	14	0,70 a 0,80
Peltophorum dubium	Canafistula	19.6	7	0,75 a 0,90
Dalbergia nigra	Bahia Jacaranda	20.8	14	0,75 a 1,22
Cariniana legalis	Jequitibá-rosa	21.7	14	0,50 a 0,65
Cordia trichotoma	Brown laurel	23	13	0,57 a 0,78
Zeyheria tuberculosa	Ipê-felpudo	24	4	0,75 a 0,80
Araucaria angustifolia	Paraná pine	27	20	0,50 a 0,61

Schefflera morototoni	Mandiocão / morotó	33	8	0,53 a 0,60
Joannesia princeps	Boleira	40	20	0,40 a 0,55

Source: CARVALHO (n.d.).

3.5.2 Adaptation

The adaptability of a species to a region is related to: climate (sunshine, frost and water deficit); soil (depth, structure and fertility); and plant health (occurrence of pests such as leaf-cutting ants, yellow beetles, sirex, aphids, defoliator caterpillars) and diseases caused by biological or physiological agents (rust, gummosis, canker, shoot blight, armillaria).

The adaptation of *Eucalyptus* species must take into account the following factors according to Ferreira (1990): bioclimatic region, climate and localities, altitude (m), average annual temperature (ºC), average annual rainfall (mm), distribution, frost (days/year), minimum ambient temperature (ºC), water deficiency and native vegetation. There are species of the genus that cannot withstand frost (*E. grandis, E. citriodora, E. urophylla*), others that are more resistant (*E. saligna, E. dunnii, E. benthamii, E. camaldulensis*), and some that are resistant to drought (*E. camaldulensis*; *E. paniculata*; *E. citriodora*). Ferreira (1979) makes recommendations for the species of the genus for each case, which can be found at: <https://www.ipef.br/publicacoes/ctecnica/nr047.pdf>.

The silviculturist's first step should be to test species and provenances that are suitable for the production objectives in the region to determine which ones are suitable. The second step is trial plantings, followed by genetic improvement of the selected species and provenances.

3.5.3 PRODUCTIVITY

There's no point in having trees of species that are suitable for the production objectives if they don't show adequate productivity and provide good profitability.

In general, *Pinus* and *Eucalyptus* species show high productivity and good profitability throughout Brazil, but other species that are being introduced still need tests to determine their growth and profitability. Teak is being cultivated in areas of the Midwest and shows good profitability; some plantations of Australian cedar and African mahogany are already showing positive results in the Southeast, but they don't yet have a consolidated market for their wood and outside these regions, where they are being cultivated, there is no information on their growth and viability. Little is known about native species, except for a few rare studies on Paricá (*Schizolobium parahyba*) and Araucaria (*Araucaria angustifolia*).

The initial productivity of *Eucalyptus* plantations, back in the 1970s, was no more than 25 $m^3.ha^{-1}.year^{-1}$ and today the average is over 35 $m^3.ha^{-1}.year^{-1}$, while some companies are already reaching 60 $m^3.ha^{-1}.year^{-1}$ in 7-year *Eucalyptus* rotations in some commercial plantations (SNIF, 2018). In 2016, Brazil reached a productivity of 35.7 m^3/ha per year for *Eucalyptus* plantations and 30.5 m^3/ha per year for *Pinus* plantations (Figura 7).

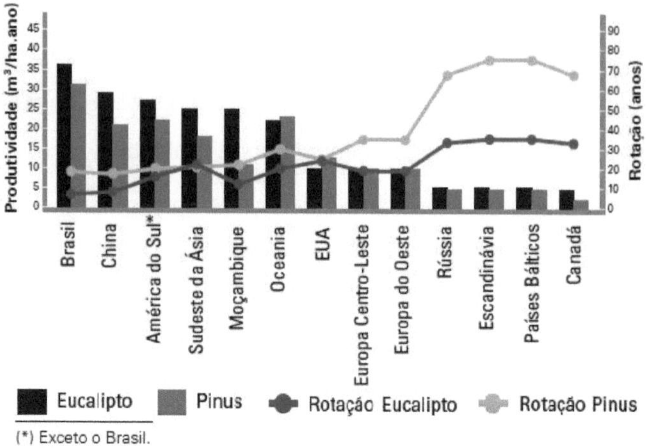

Figura 7 - Productivity and average rotation in different regions. Source: IBA (2017).

3.6 Statistics

The most common forest statistics used to describe forest structure are measures of central tendency, dispersion and relationships between variables. After describing the structure, we move on to studying growth and then modeling trees and stands, in order to finally predict growth and production, usually by site class.

The main statistics used to describe forests and correlations between variables are listed in Tabela 8 .

Tabela 8 - Statistics commonly used to describe forests and relationships between variables.

Type	Statistics	Observations	Equation
Central tendency	Arithmetic mean ($\bar{X}$)	It is obtained by adding the observed values divided by the number of elements	$\bar{X} = \frac{\sum_{i=1}^{n} X_i}{n}$
	Quadratic mean (*Xq*)	It is obtained by the square root of the sum of the square of the observed values, divided by the number of elements.	$X_q = \sqrt{\frac{\sum_{i=1}^{n} X_i^2}{n}}$
	Weighted average (*Xp*)	It is obtained by assigning weights (p_i) which can be proportions of occurrence, proportion of total area, etc.	$X_p = \left(\frac{\sum_{i=1}^{n} p_i . X_i^2}{\sum_{i=1}^{n} p_i}\right)$
	Median (ME)	It is the element that divides the population into two parts with the same number of elements.	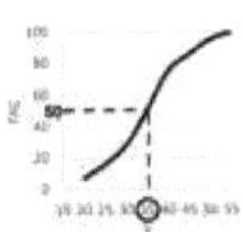

Type	Statistics	Observations	Equation
	Fashion (MO)	It is the element(s) with the highest frequency in the population.	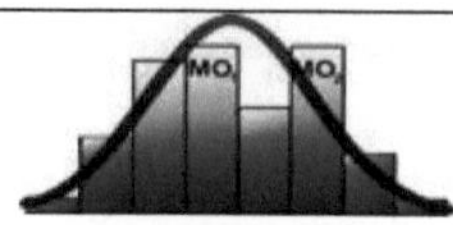
Dispersion	Range of variation	Amplitude (AV) is the difference between the highest (LS=upper limit=Xn) and lowest value (LI=upper limit=X1) of a variable X	$AV = LS - LI = X_n - X_i$
	Variance (S^2)	It is the mean square of the deviations from the mean.	$S^2 = \frac{\sum_{i=1}^{n}(X_i - \overline{X})^2}{n-1}$
	Standard deviation (S)	It is the square root of the variance.	$S = \sqrt{S^2}$
	Coefficient of variation (CV)	It is the expression of the standard deviation as a percentage of the mean.	$CV = \frac{100.S}{\overline{X}}$
Correlation between variables	Pearson's correlation	Used for data with a normal distribution.	$r_{xy} = \frac{\sum_{i=1}^{n}(x_i - \bar{x})(y_i - \bar{y})}{\sqrt{\sum_{i=1}^{n}(x_i - \bar{x})^2 \sum_{i=1}^{n}(y_i - \bar{y})}}$
	Spearman's rank correlation	Used for data without a normal distribution.	$r_s = 1 - \frac{6 \cdot \sum_{i=1}^{n} d_i^2}{n^3 - n}$

Source: Wonnacott and Wonnacott (1980).

3.7 Characterization of trees and stands

Forests are characterized by dendrometric variables, space, time, forest community and forest stand structure.

- Dendrometric characteristics - diameter, height, basal area, volume, trunk shape and tree crowns;
- Space: location, area occupied and spaces influenced by forestry, industrial and commercial activities resulting from the forestry enterprise;
- Time: stand age, cycles and forest rotation;
- Forest community: species that make up commercial stands and other forest areas;
- Forest structure: density, frequency and dominance of species; species richness and diversity, etc.

3.7.1 Diameter

It is one of the most important variables for forest management. Tree diameter is defined as the diameter taken at breast height, which varies from country to country, being taken at 1.3 m from the base of the tree in Brazil, 1.37 m in the United States, 1.29 m in England and other European countries and 1.25 m in Japan.

Diameter is the variable that is always taken from all the trees in the sampling units in forest inventories, as it is easy to measure and is closely related to the other tree variables, such as height, trunk volume and crown diameter. The unit of measurement for diameter is the centimeter and it is generally represented to one decimal place.

Diameter expressions

Diameter can have different expressions, the main ones have the following definitions:

- Arithmetic mean diameter ($\bar{d}$) - is the arithmetic mean of the diameters;
- Mean basal area tree diameter (d_g) - is the quadratic mean of the diameters; can be obtained by $d_g = \sqrt{\bar{d}^2 + S^2}$;
- Weise diameter (d_w) - is the diameter located at a cumulative probability of 0.6 and approximates the tree to the average volume of the stand;
- Average diameter of the central basal area tree (d_z) - is the diameter obtained at half the accumulated basal area of the stand;
- Hohenadl diameters (d_- , d_+) - are the diameters obtained by decreasing and adding a standard deviation (s) to the mean (d): $d_- = \bar{d}$-s; $d_+ = \bar{d}$+s.

The relationship between the expressions for the diameter of an equiannual forest are as follows:

$$d- < \bar{d} < dg < dw < dz < d+$$

DIAMETER DISTRIBUTION

Statistical summaries are usually carried out by classifying diameters into size classes and studying their distribution. The distribution of diameters is analyzed in diameter classes with intervals calculated by any method such as Sturges' rule ($k = 1 + 3.3 \log_{10} n$), or Floriano's rule ($k = n^{0,175} . \Delta d^{0,3}$), where k is the number of classes to be used, n is the number of observations (trees) and Δd is the range of tree diameters.

In planted forests, the distribution of diameters is generally close to the normal distribution. However, when there are irregularities in the environment, or after thinning, the normal distribution may no longer fit the characteristics of the forest. In these cases, other types of models are used, such as the Weibull, Beta, Gamma and Lognormal distributions, among others (see 'section 4.8). A Figura 8 shows the fitting of probability density distribution functions for the diameters of a candeia (*Eremanthus erythropappus*) plantation, carried out by Ribeiro et al (2014).

For inhospitable forests, negative exponential models are used, or cumulative distribution models, which are generally more appropriate in this case. In natural forests, the distribution is close to a negative exponential, but there is usually a deformation in the curve, as young trees when they reach maturity are more resistant to diseases, pests and bad weather and remain healthy for a long time until they enter senility, when mortality increases again.

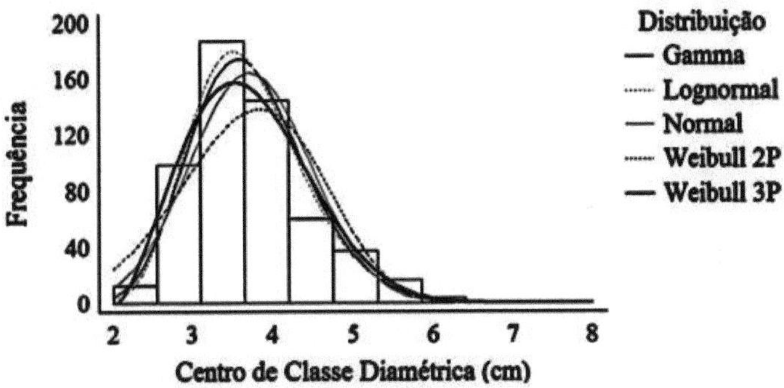

Figura 8 - Diametric distribution and probability density functions adjusted for an experimental candeia plantation. Source: Ribeiro et al (2014).

One example is the distribution of andirobe trees found by Abreu et al (2014) in Amapá (Figura 9), where the estimation errors in relation to the observations can be perfectly seen by the adjusted negative exponential equation.

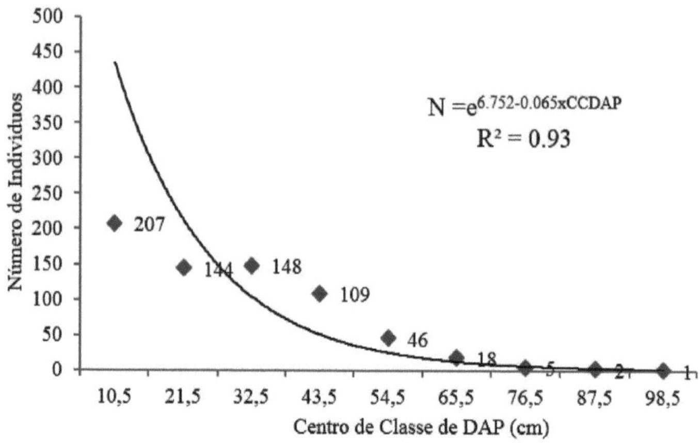

Figura 9 - Frequency distribution of andirobe trees in an area of the Fazendinha APA, Macapá. Source: Abreu et al (2014).

3.7.2 Height

The main statistical expressions of tree height in a forest stand are as follows:

- Arithmetic mean height ($\bar{h}$) - is the arithmetic mean of the tree heights;
- Hohenadl tree heights (h_- , h_+) - these are the heights of Hohenadl trees with d_- and d_+ , estimated using the hypsometric relationship [d=f(h)];
- Height of the tree with average basal area (h_g) - is the height of the tree with the diameter of the tree with average basal area, obtained from the hypsometric relationship;
- Weise tree height (hw) - is the height corresponding to the tree with the Weiss mean diameter, obtained from the hypsometric relationship;
- Height of the tree with the central basal area (hz) - this is the height of the tree with the diameter corresponding to the tree with the central basal area, obtained from the hypsometric relationship.

- Lorey height (h_L) - is the height obtained by weighting the heights with the basal area; in inventories using the Bitterlich method it is the arithmetic average height of the trees in the sampling unit, as the height is already weighted by the basal area in this type of sampling; it can be calculated for fixed area sampling units using the equation:

$$\square_L = (\sum_{i=1}^{n} gi.\square i)/\sum_{i=1}^{n} gi;$$

- Assmann's dominant height (h_{100}) - is the average height of the 100 thickest trees per hectare;
- Weise dominant height (h_0) - is the average height of the thickest 8% of trees in the stand per hectare.

3.7.3 Basal Area

The individual basal area (g) of a tree is determined by the square of the diameter (d) multiplied by the ratio of π/4 ($g=\pi*d^2/4$), i.e. the cross-sectional area of the trunk at breast level (1.3 m height). The basal area of the stand (G) is defined as the sum of the individual basal areas of the trees per hectare. The unit of measurement for basal area is the square meter and, when expressed for the stand, it is given in square meters per hectare.

The basal area of stands (G) is important for forestry because, like diameter, it is correlated with trunk volume and competition between trees and can be used to control stand density when associated with the age and quality of sites. The basal area of forests tends to increase with age as trees grow; older stands support greater basal area and competition between trees than young stands. Basal area per hectare also varies with tree density. In young stands, or with very low stocking density, the basal area can be close to zero, while in old stands, with high population density, it can exceed 100 m²/ha (WEST, 2009).

In Figura 10 curves for density control of *Pinus elliottii* in the Serra do Sudeste, Rio Grande do Sul, are plotted as a function of minimum and maximum basal area per hectare, by age and site index; the site indices refer to the dominant height at the reference age of 22 years (FLORIANO, 2006).

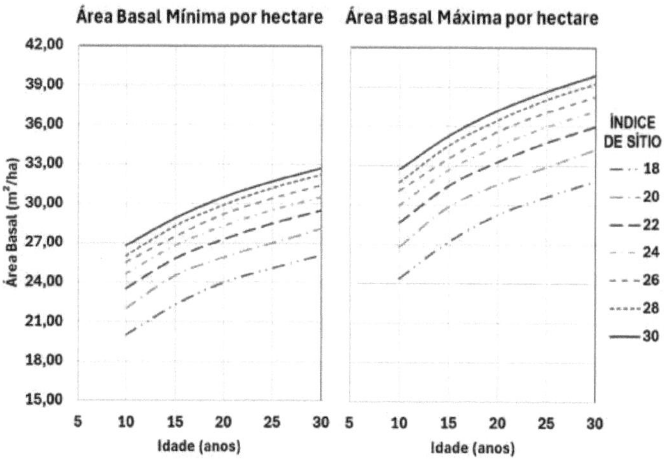

Figura 10 - Minimum and maximum basal area curves per hectare for *Pinus elliottii* in the Serra do Sudeste do RS.

3.7.4 Volume

The volume of wood in the tree trunk is the main commercial product of cultivated forests and contains a significant amount of the total biomass.

Volume is the result of the accumulated increase over a given period of time, the amount of which depends on the species, age, site, rotation and management system used.

The volume of production units is the most important variable for production planning, which is why management is interested in the following types of volume: actual volume, normal volume, ideal volume and desired volume.

Actual volume

This is the stock of wood that actually exists in the stand, determined using a sampling procedure, subject to a tolerable sampling error.

Normal volume (Ecological Forests)

It is the volume determined for a management class, which is conditioned by the similarity of species composition, site, same rotation and management system, and is obtained based on the normal forest model.

Ideal volume (Inequitable Forests)

This is the volume that best corresponds to the conditions of the site and stand. It is the volume determined to be ideal for the management of inhospitable forests, obtained according to the balanced forest model.

Desired volume

This is the volume defined for certain production planning objectives. It is usually determined on the basis of the ideal distribution of stands into age classes.

3.7.5 Tree shapes

The main components of trees are the root, trunk and crown and they can take many forms (Figura 11).

Figura 11 - Different forms of large plants. Source: Gennia News (2018).

ROOT SHAPE

The roots of trees and pines generally have a pivoting root that is not very long and forms a cup with the ends of the roots, but this will depend on the characteristics of the species, fertile areas and the resistance of the ground. Palm trees, which can also be considered trees, have a fasciculated root. The study of roots is carried out by digging trenches in radii from the trunk and by sampling, drawing and weighing the roots in the trench and measuring the volume of the trench, it is possible to draw a profile and estimate the biomass of roots of the whole tree.

TREE CANOPY

Trees that grow free from competition have a wider crown and a thicker trunk at the base and are generally lower, which is called the forest form, while trees that grow inside the forest have a longer and narrower crown, a more cylindrical trunk and greater height, which is called the forest form. Depending on the type of growth, they may have a single dominant apical bud, forming longer, pyramidal or elongated crowns, which are called monopodial (with apical dominance); others have no dominant bud in the crown and the branches all grow in the same way, forming multiple branches of the same size, giving a more rounded shape, which is called sympodial (without apical dominance) (Figura 12).

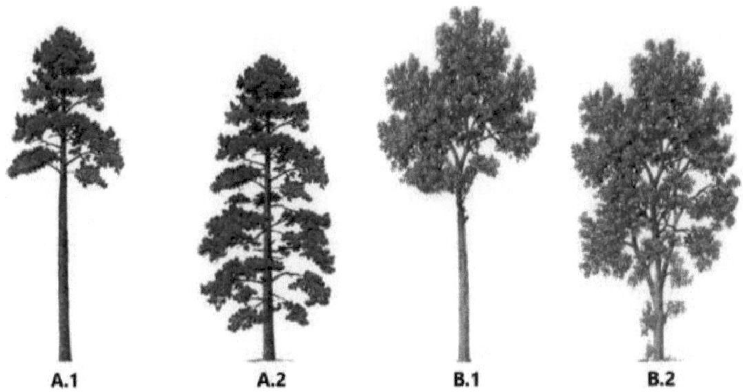

Figura 12 - A) Monopodial growth, forest form (A.1), natural form (A.2); B) Sympodial growth, forest form (B.1), natural form (B.2). Source: Imaña et al (2002).

CANOPY MEASUREMENTS

The tree's canopy is responsible for photosynthesis and therefore for its growth. The more a tree photosynthesizes, the more it grows. The intensity of photosynthesis is related to the size of the canopy, especially the area directly hit by sunlight, and the genetic potential of the trees.

The different shapes that canopies can take are shown in Figura 13 .

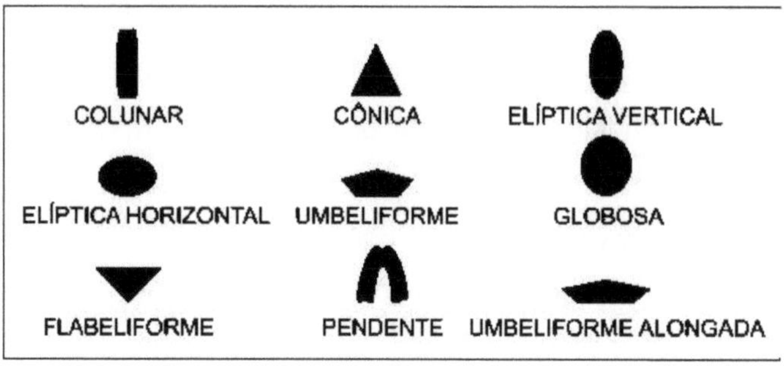

Figura 13 - Shapes of the treetops. Source: Copel (2018).

When it comes to forest stands, the more the crown surface covers the ground surface, the higher the productivity per unit area of land. Therefore, by measuring the diameters of the tree crowns and calculating the surface area of the crowns over the ground, it is possible to estimate the percentage of ground cover.

Crown density is also used as an index of stand density, making it possible to estimate the need for thinning.

The crown projection surface (SPC) of a tree is determined by measuring four crown radii in the north (RN), east (RL), south (RS) and west (RO) directions, and the projection surface is calculated by:

$$SPC = (\pi/4) \, . \, [(RN+RL+RS+RO)/2]^2$$

Some authors suggest measuring a greater number of rays from the treetops, but this is only necessary when you want to study them individually, in which case you may need between 8 and 16 rays (LAAR and AKÇA, 2007; ASSMANN, 1970). For modeling or statistical calculations, four radii are usually sufficient, and it is recommended to measure the radii in the north, east, south and west directions, due to the different insolation in each direction (Figura 14).

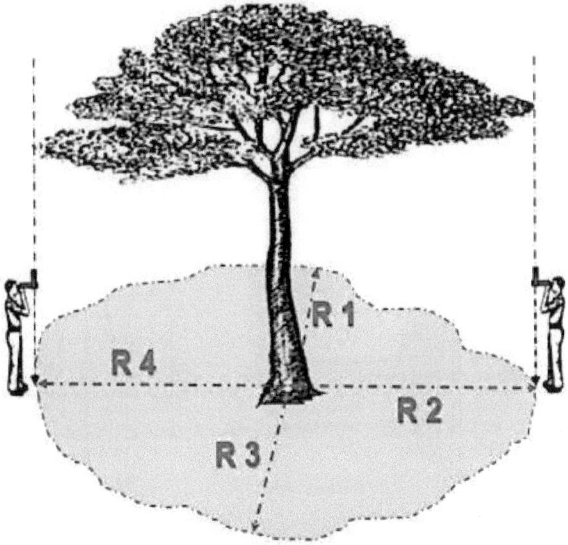

Figura 14 - Measurement of crown radii to determine the projection surface above the ground.

The total surface area of the crown is estimated by calculating the surface area of solids of revolution to which the shape of the crown of the individuals resembles, and it is necessary to calculate the integral from the top to the base of the crown to obtain the surface area. The same applies to estimates of crown volume.

According to Laar and Akça (2007) canopy characteristics are useful for predicting growth. In spacing, thinning and fertilization trials, the crown can be correlated with growth and soil water availability, where the close relationship between crown size and the amount of photosynthetically active foliage can be seen. The increase in crown radius with increasing tree age is mainly controlled by competition from neighboring trees.

In dense positions, the eccentricity of the crown is quite high, due to unequal competition from the surrounding trees. For this reason, the crown radius is usually measured in 4 directions or more. The general morphology of a tree's crown is illustrated in Figura 15 .

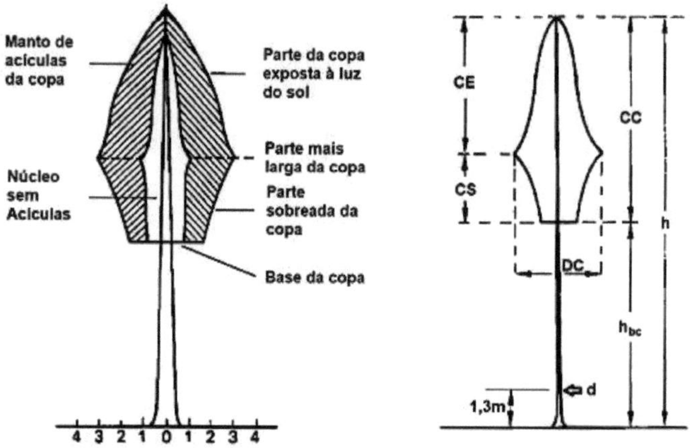

Where: CE = crown length in sunlight; CS = crown length in shade; DC = crown base width = crown diameter; CC = crown length; h_{bc} = crown base height; h = tree height; d = tree diameter.

Figura 15 - Morphology of the *Picea* canopy. Sources: Burger (1939), Assmann (1970).

Morphometry

Tree morphometry can be defined as the study of tree shape through crown dimensions, tree diameter and height and the relationships between these variables. The relationships between trunk diameter, tree height and crown dimensions, as shown in Figura 16 are useful for assessing competition between individuals, forecasting growth and predicting the need for silvicultural treatments.

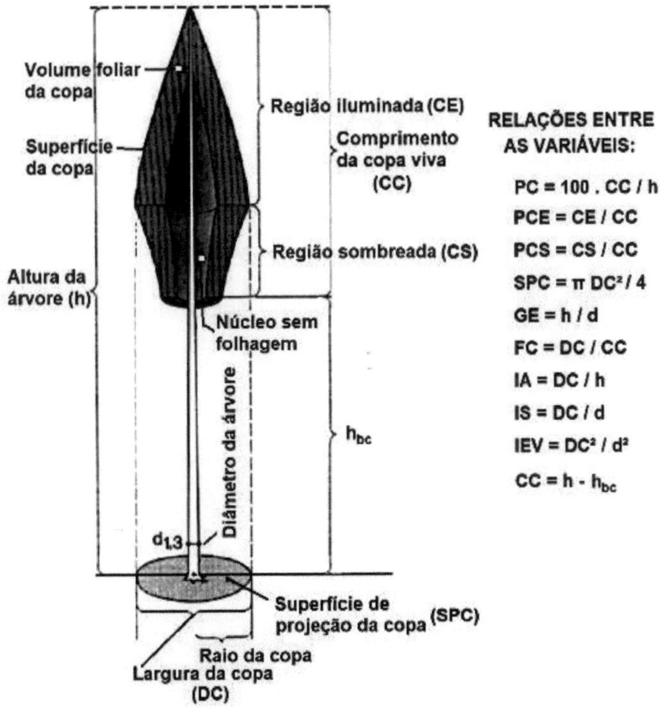

Figura 16 - Morphology and morphometric relationships of the *Picea abies* crown. Source: Burger (1939); Assmann (1970); Laar and Akça (2007).

The relationships between the variables make it possible to investigate the situation of competition and tree growth.

Crown length is determined by the total height of the tree (h) subtracting the height of the base of the crown (h_{bc}). Care must be taken when determining the height of the crown base, which is from the ground to the start of the green crown. This height is sometimes confused with the height of the stem (h_f), but the concepts are different, as the height of the stem is the height of the commercial part of the trunk and has nothing to do with the crown.

Crown percentage (PC = 100 . CC / h) shows how much of the height is covered by living crown. Floriano (2004) studied the effect of artificial pruning on Pinus elliottii and showed that the effect of crown reduction begins to be felt when more than 40% of the tree's total height is removed, and is already quite significant when 50% is removed.

The Proportion of Canopy Exposed to Sunlight (PCE = CE / CC) is the part of the canopy that is responsible for most of the photosynthesis carried out by the tree and therefore for its growth. While the shaded part may even consume more nutrients than it produces. Many species tend to lose their lower branches when they become shaded by the rest. When the Proportion of Shaded Canopy (PCS = CS / CC) is high, it can be an indication that there is a lot of lateral competition. It can also be interpreted as a need for pruning. The more elongated the crown, the greater its surface area for the same projection area on the ground and, therefore, the more light the crown receives, making the tree more efficient at taking advantage of the land area.

The Crown Projection Surface (CPS = $\pi DC^2 / 4$) is used to assess ground cover. While the trees are growing free of competition, the canopies are wide and as they begin to touch and shade each other, the canopies become narrower and individual growth is affected, being reduced the smaller the tree in relation to the others, up to the extreme degree of competition with the death of the smaller trees, which is called self-thinning. According to Assmann (1970), the average number of trees per hectare (N) can be determined as a function of the space occupied by the tops of trees growing free of competition, as follows:

- In square spacing: $N = 10000 / DC^2$;
- In triangular (or hexagonal) spacing: $10000 / (0.866.DC^2)$;
- In mixed spacing (between square and hexagonal): $10000 / (0.933.DC^2)$.

The crown projection surface (CPS) of an individual tree can be referred to as approximately proportional to the growth space or nominal area occupied by this tree (SEEBACH *apud* Assmann, 1970).

The Degree of Slenderness (GE = h / d) is an indicator of the cylindricity of the trunk; the higher it is, the greater the cylindricity, which may indicate excessive population density over long periods.

The crown shape (FC = DC / CC) is the corresponding crown shape to the degree of slenderness of the trunk. More elongated canopies are related to greater competition between trees, but if it is a natural characteristic of the species, it can also be an indicator that the species is more efficient at occupying space, as a more elongated canopy receives more light than a more rounded one.

The Coverage Index (AI = DC / h) is an indicator of space occupation as a function of height; larger canopies mean that the trees have little competition with each other.

The Overhang Index: (IS = DC / d), also known as the Seebach Growth Space Factor, or crown projection ratio, shows how much the crowns are contributing to the thickening of the tree trunk; the lower the IS, the more the trunk is thickening with a smaller crown diameter.

The Living Space Index ($LVI = DC^2 / d^2$) is also known as the ground cover area quotient and shows how much living space is occupied by the tree in relation to its basal area. The study of minimum and maximum VSI limits by species and site class makes it possible to determine the minimum and maximum surface area to be occupied by a tree (ASSMANN, 1970).

Cup standings

A classification system for tree canopies in relation to their position in the forest canopy is useful when discussing stand development, such as the one below:

1) Emergent - These are the trees that project their crowns well above the canopy, even above the dominant trees; they receive full light from above and from the sides; their occurrence is very small and they are usually selected for genetic improvement programs;
2) Dominant - Trees with a crown extending above the canopy that receive full sun from above and partly from the sides. The crowns are well developed and wider than the canopy;
3) Codominant - These are trees with canopies that form the level of the canopy and receive full light from above, but little from the sides; their canopies have a medium radius;
4) Intermediate - These trees are lower than those in the previous classes, have a smaller crown diameter

and their tops receive direct light, but none from the sides;

5) Overtopped - These are trees whose crowns are below the level of the canopy and do not receive direct light.
6) Lobas - These are trees that develop and grow free from competition; they have branches on all sides, with branches well below the level of the canopy; their canopies receive direct light from above and on most of the sides.
7) Dead - These are dead trees, regardless of their position in the canopy; suppressed trees tend to die from excessive competition and some of the other classes may die from disease or pest attack, or even from being struck by natural phenomena such as lightning.

3.7.6 Trunk shape

The trunks of trees can have different shapes, from the most regular, such as the *Araucaria* (*Araucaria angustifolia*), to the crooked, such as the Mangabeira (*Hanconia speciosa*) (Figura 17). The most important part of the trunk is the stem, which is the commercial part of the tree trunk and is the main object of study for quantifying commercial wood (ORMOND et al., 2006).

Figura 17 - ***Araucaria angustifolia*** **(A) and** ***Hancornia speciosa*** **(B). Source s: A) author; B) Lima (2010).**

The study of the shape of the trunk is usually associated with solids of revolution such as cylindrical, parabolic, conical or neiloid, which can be carried out using the following function (GARCÍA, 2004):

$$g_x = \text{p.x}^r$$

$$v = \int_{x=0}^{h} g_x \, . \, d_x$$

Where: v = volume of the solid of revolution; g_x = cross-sectional area of the trunk at distance x from the top (m²); x = distance (m) from the top; p, r = coefficients, where p is the parameter that represents size, r is the parameter that represents shape; h = height of the tree.

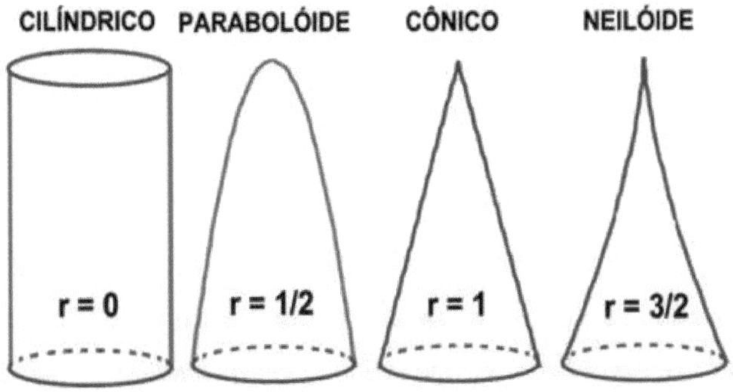

Figura 18 - Geometric shapes that tree trunks resemble and the value of the parameter (r) of the function g_x =p.x .^{r}

ARTIFICIAL FORM FACTOR (F)$_{1,3}$

The artificial shape factor ($f_{1,3}$) provides a rough estimate of tree volume and is determined by the ratio between the actual tree volume and a cylinder with a diameter equal to the tree diameter (d) and the same height (h):

$$f_{1,3} = (real\ volume) / (h \cdot \pi \cdot d^2 / 4)$$

The volume of the tree can be calculated by:

$$v = f_{1,3} \cdot h \cdot \pi \cdot d^2 / 4$$

The form factor should be determined by the stand's diametric class and density, as the shape of the trunk varies with the size of the trees and the competition between them.

NATURAL FORM FACTOR (F)$_{0,9}$

In this case, the tree diameter considered is $d_{0,9}$ from the Hohenadl method and the shape factor ($f_{0,9}$) is calculated using the formula:

$$f_{0,9} = actual\ Hohenadl\ volume\ (v) / cylinder\ volume\ (vc)$$

The natural form factor can be found from the equation:

$$f_{0,9} = v / (h \cdot \pi \cdot d_{0,9}{}^2 / 4)$$

Or, the volume of the tree can be calculated by:

$$v = f_{0,9} \cdot h \cdot \pi \cdot d_{0,9}{}^2 / 4$$

SHAPE QUOTIENT

The shape quotient is the ratio between the diameter at a given height (d_i) and the diameter of the trunk at 1.3 m from the ground (DAP). It has no unit of measurement and is expressed as a proportion, with >0 and usually <2. It is calculated using the function:

$$q_{0,ih} = d_{,0ih} / DAP$$

Where: $q_{0,ih}$ = shape quotient at 0.i times height; $d_{0,ih}$ = tree diameter at 0.i times height; DBH = diameter at breast height, taken at 1.3 m from the base of the trunk; 0.i = proportion of the diameter measurement height.

The shape quotient can be greater than 1 when measuring diameters below the DBH, or in trees with unusually shaped trunks, such as *Ceiba speciosa*, or even with deformations (Figura 19).

Figura 19 - *Ceiba speciosa*. Source: Viale (2023).

Artificial shape quotient ($q_{0,5h}$)

It is calculated by measuring the diameter at half the total height ($d_{0,5}$) from the base to the top of the tree and dividing by the diameter at 1.3 m height (DAP).

$$q_{0,5h} = d_{0,5h} / DAP$$

Natural form quotient ($\eta_{0,5h}$)

This is the ratio between the diameter at 50% height ($d_{0,5h}$) and the diameter at 90% height ($d_{0,9h}$), from the top of the tree, i.e. with the Hohenadl diameters $d_{0,1}$ and $d_{0,9}$, as shown in the equation

$$_{0,5h}\eta = d_{0,1h} / d_{0,9h}$$

Hohenadl shape quotient (q_H)

It is the shape quotient defined by the ratio between the diameter at 1.3 m from the base (DAP) and the diameter at 90% of the height from the top ($d_{0,9h}$):

$$q_H = DAP / d_{0.9h}$$

3.7.7 Biomass

Biomass is measured separately for the trunk, branches, foliage and roots.

The trunk can be measured using any cubing equation, both with and without bark, and its density determined by drying it in an oven and then weighing representative samples taken along the trunk, determining its density and then the total mass of the trunk.

The branches are usually weighed directly when they are still green and then representative samples are taken, which can be proportional to the level of branching, and then dried and weighed to determine the density and then the total mass of branches in the tree.

The leaves are subjected to a process similar to that of the branches, weighing the total, or a proportional sample, and then subjected to drying and weighing to determine the density and then the mass of the tree's leaves.

Roots are more complicated to measure because they are in the forest substrate. One way of estimating their biomass is to fell the tree and open a trench taking in the entire diameter of the maximum projection of the tree's crown. All the roots in the trench should be sorted by size, cleaned, weighed and dried while they are still green so that they can be weighed and the total mass determined later as a function of the proportion of the volume of the trench in relation to the volume calculated using the surface projection of the crown and the depth of the trench.

3.7.8 Mortality

Mortality (M) differs between equiaquian planted forests and natural forests that are not equiaquian. In planted forests, initial mortality is higher and decreases until it practically stabilizes, generally becoming rare due to the care taken in silviculture; therefore, it occurs differently by age class of the forest stands, being higher in the smaller age classes. In uneven-aged natural forests, mortality is differentiated by diameter class, and is higher in the smaller diameter classes. Thus, modeling mortality (M) in equiaquatic forests should have the age (t) of the trees as the main independent variable [M=f(t)], but others can be added, such as the site index (IS), representing the functions by [M=f(t, IS)]. In inhospitable forests, mortality modeling should have diameter (d) as the main variable, represented by the function [M=f(d)], which can be modeled globally for all species, or particularly for species of interest.

3.7.9 Wood quality

The quality of the wood is a fundamental part of forestry production planning, even when it is considered a secondary product, as in the case of the production of bark for the manufacture of tannin from black wattle stands, since it can represent an increase in the value of the product, or greater ease of sale. In the case of acacia stands, a higher density and calorific value of the wood could be important at the time of sale, but this is overlooked in plantations in southern Brazil. Higher wood density could be achieved through proper management of stands and genetic improvement, for example.

The definition of the quality of the wood to be obtained must take into account the most desirable assortment or the one of greatest value to the producer. In some cases, a higher basic density is desirable, so tighter spacing is used and the stands are managed at a higher density to produce a greater quantity of autumn-winter wood (denser) than spring-summer wood (less dense). In other cases, it's the volume that matters, regardless of the size of the wood, and high planting densities are used. However, if the need is for larger wood in greater volume, the initial spacing should be greater and the density management should be carried out with more space for the trees to grow individually.

Another factor that affects the quality of the wood, depending on the objective, is pruning, both natural and artificial. If you want wood with fewer knots and smaller dimensions, tighter initial spacing and higher stand density management are recommended, as the branches will be thinner and natural pruning will be more intense. However, artificial pruning generally has better results and can be combined with wider spacing and lower stand density management to obtain larger timber in shorter rotations, resulting in lower accumulated interest on stand costs and consequently higher profits.

Fertility and other soil characteristics have a major influence on growth. Fertilization provides greater growth and larger wood more quickly, but generally has a lower basic density, which greatly influences the quality of the wood.

The climate is another factor that influences the quality of the wood, because in warmer and rainier climates, the trees grow more.

In short, all the environmental factors, management and genetic characteristics of the trees influence the quality of the wood.

Thus, the main factors influencing wood quality that can be partially or completely controlled by man are:

- germplasm;

- the soil - moisture, compaction, fertility;
- planting density;
- managing the density of stands;
- artificial pruning.

3.8 Forest space

It is represented by the company's various territorial units and the space in which its market is located. The forest space is characterized by the areas covered by natural and cultivated forests and other spaces associated with them, such as areas covered by other types of vegetation, areas dedicated to infrastructure, the surface of water slides, etc.

3.8.1 Preservation and conservation areas

The forest manager's first step in managing the forest space is to identify the water slides and springs. Subsequently, in accordance with Law 12.651/2012 on the protection of native vegetation, it is necessary to identify the permanent protection areas and locate the legal reserve areas, defined in the Law as:

- Permanent Preservation Area - APP: a protected area, whether or not covered by native vegetation, with the environmental function of preserving water resources, the landscape, geological stability and biodiversity, facilitating the gene flow of fauna and flora, protecting the soil and ensuring the well-being of human populations;
- Legal Reserve: an area located within a rural property or possession, delimited under the terms of art. 12, with the function of ensuring the sustainable economic use of the rural property's natural resources, assisting the conservation and

rehabilitation of ecological processes and promoting the conservation of biodiversity, as well as the shelter and protection of wild fauna and native flora.

Next, it is essential to locate support points for forest security, such as water reservoirs and observation towers for fire detection. The next stage is to lay out the forest roads and then divide up the production units and create an identification system for them. Both the productive areas and the conservation or preservation areas are costly for the company and therefore need some kind of accounting identification, with cost centers that make it possible to monitor the economic results of each unit in order to facilitate its management. Cost centers represent each autonomous unit of a company that has autonomy in terms of financial management (income and expenses), and the combined results of all the cost centers will represent the results of the entire company.

3.8.2 Spatial divisions and subdivisions

Areas can be divided into ecological, productive and technical classifications. The classification of areas is defined particularly by each company, but in general they can be defined as follows:

- Ecological - classification of areas according to habitat;
- Productive - is the technical and silvicultural classification in a hierarchical manner into regions, districts or gardens, farms, management units, plots, sections and subsections. Production units should preferably have a standardized size and

shape; standardized use and strict statistical control (stock, cutting and planting).

Units are generally defined in up to 5 levels, as follows:

1) Region: an area of national territory that concentrates the company's properties under the same management;
2) Districts or hortos: these are smaller subdivisions of regions that concentrate nearby farms;
3) Farm: these are usually contiguous areas registered with a notary;
4) Management unit - these are continuous production areas, with or without subdivisions such as plots and sections, with plantations usually of the same species and age, under the same silvicultural system; they are the units that receive an accounting cost center; they can bring together several plots, not necessarily contiguous, but close, with similar environmental characteristics; they are the main units for forest control, planning and management.

Divisions and subdivisions should be given a hierarchical numerical code to facilitate their control and identification, and management units can always be identified by their accounting cost center code in order to integrate forest management with business accounting.

TECHNICAL SUBDIVISION OF FOREST AREAS

It is carried out by utilization classes (areas of similar and close stands of the same species and rotation with the same economic goal). It is carried out with the aim of controlling the ordering of stands, summarizing information on the different site and age classes, planning and controlling the sustainability of production for the different classes.

3.8.3 Road network

The use of rural spaces depends on access, so the location of access routes, especially to productive areas, comes next. According to Carvalho (2013): "obligatorily, the roads must meet the activities of forestry, forest protection, small activities, neighbors, third parties, harvesting and transport". Initially, the mandatory crossing points and places to avoid are located. Obligatory points are the places where bridges and culverts can be built, access points to neighbors, water supply sites (reservoirs, rivers, etc.), access to observation towers, among others. Areas to avoid are places where the construction of the road could affect the areas to be preserved and places where the cost of construction is very expensive, such as very steep slopes, rocks that need to be blasted, marshes, streams and torrential rivers.

The road network is expensive to build, so it must be optimized by building the minimum to obtain the maximum transport capacity. One of the ways of designing forest roads is the "herringbone" (Figura 20), made up of parallel roads that run at an angle of around 60º to a main road, which in turn runs into other more important roads in the same way.

The angle of insertion must be sufficient to allow maneuvering and access for the extra-heavy trucks currently used to transport timber. At the end points of the secondary road network, provision must be made for trucks to turn around and for parking areas, loading timber and unloading materials, inputs and machinery used in forestry, maintenance and timber harvesting.

The time of laying the roads can be carried out as recommended by Carvalho (2016), as shown in Figura 21 .

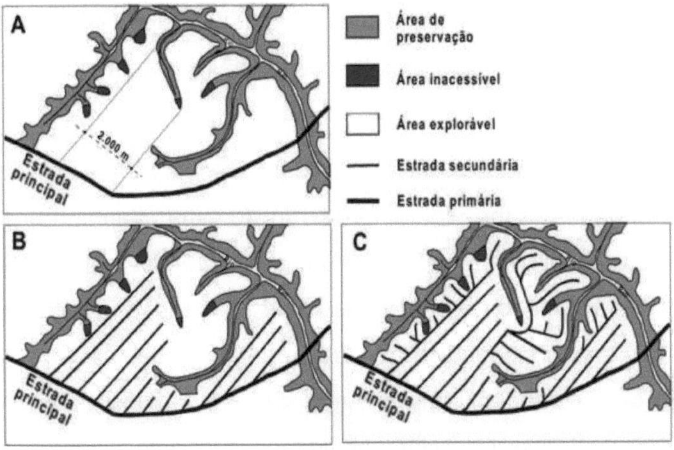

Figura 20 - Herringbone planning of forest roads. A) Main road; B) Secondary roads; C) branches of secondary roads. Source: IAMAZON (2015).

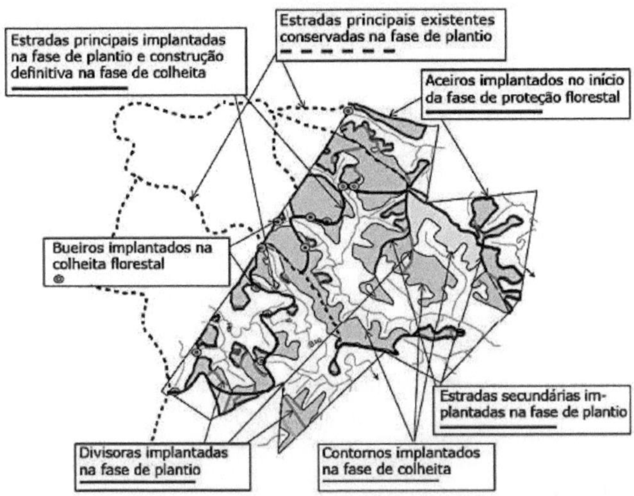

Figura 21 - Time of implementation of forest roads. Source: Carvalho (2016).

The cost of forest transportation is mainly related to the cost of building the road itself, its maintenance, the cost of extraction (hauling) and the reduction in the productive area (Figura 22), and also with the cost of the main transportation to the market or industry.

After planning the roads, it is necessary to locate the boundaries of the plots intended for forestry, planning how the wood will be transported from inside the plot to the edge of the access roads where the wood will be loaded onto the main transportation vehicles (trucks). Plots should preferably be of a size that minimizes the cost of transporting the wood from inside the plot to the unloading and loading sites.

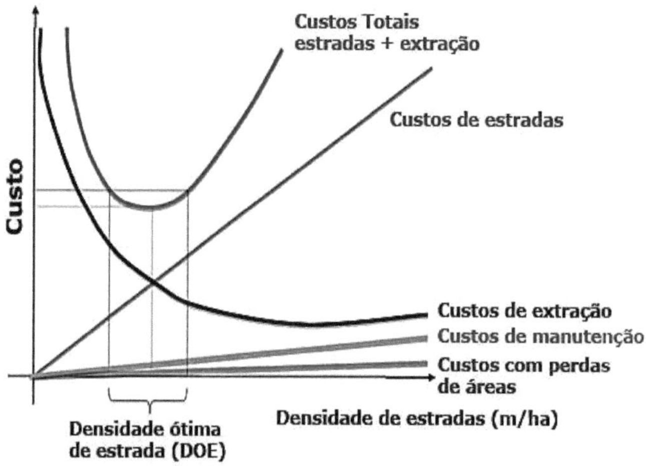

Figura 22 - Determining the optimum road density according to Carvalho (2016).

3.9 Time

Forest time is regulated by the age of the trees and is treated differently for natural forests and planted forests.

The most common time periods considered in forest management are:

- Rotation - Eucalypt forests - Period of time between planting and final harvest;
- Thinning cycle - Period of time between thinning;
- Felling cycle - this is the term used for inhospitable forests, meaning the time between one production cut and the next;
- Duration of production - is represented by the age of utilization; it is the time it takes for the forest to become productive, and is determined according to the productivity of the forest, the market, access routes, weather conditions and the labor available for harvesting;
- Renewal period - the period between cutting and the start of the new crop;
- Planning period - the planning period for the development plan;
- Transformation period - the period required for the transformation of a forest from one silvicultural system to another;
- Balancing period - period for balancing the shortage or excess of stock in each age class;
- Planning horizon - The period of time considered for production planning or economic evaluation of a forestry enterprise.

3.9.1 Age

In old-growth forests, it is the time elapsed from planting to the moment in question. In natural forests, there is generally no characterization of age, as this is unique to each individual.

The period of time between planting and final harvesting in planted forests is called the rotation, while the period of time between one cut and the next in natural forests is called the cutting cycle.

Stands can be classified according to their age (t) as equiannual or inequiannual:

- Equal - the age range of the trees in the stand is between +/- 10% of the average age;
- Inequitable - the age range of the trees in the stand exceeds +/- 10% of the average age.

AVERAGE AGE

Arithmetic mean age

$$I = \sum t_i / n$$

I = arithmetic mean age; ti = age of tree i; n = number of trees.

Average age of areas

$$I_a = \sum (tj \, . \, Aj) / \sum Aj$$

I_a = average age of areas; Aj = area of stand j; tj = average age of trees in stand j.

Average age of volumes

$$I_v = \sum (tj \, . \, Vj) / \sum Vj$$

I_v = average volume age; Vj = stock volume of stand j; tj = average age of stand j.

3.9.2 AGE CLASSES

Natural age classes

- Tall managed stands go through the following stages: renewal, dense, thinning, timber;
- Coppice managed stands go through the stages of: sprouting, rebound.

Silvicultural age classes

The stands are considered to be in the renewal, silvicultural treatment or improvement phase.

Artificial age classes

In this case, which is more common for forest management, stands are divided into age classes in years.

3.10 Forest structure

The forest structure is mainly characterized by:

1 - Species;

2 - Horizontal structure;

3 - Vertical structure.

Productive forest areas are interspersed with protected areas and the study of forest structure is important not only for the management of productive forests, but also for protected areas. Monitoring forest structure makes it possible to verify the maintenance of species richness, biodiversity, evolutionary status, growth, density, etc., facilitating the planning of production or actions to guarantee its conservation and preservation.

3.10.1 Species

The floristic composition of an ecosystem is represented by the list of species that exist in it, and is quantified mainly by the diversity and aggregation of species.

DIVERSITY OF SPECIES

Diversity indices are indicators of the state of ecosystems (Magurran, 1988) and estimate two basic elements: richness and evenness. Richness is the number of species in a community. Evenness is the number of individuals per species. Diversity indices can be divided into three types:

- Richness - measures the number of species in a sample unit;
- Abundance - measures the number of individuals of a species in a community;
- Proportional abundance - measures wealth and uniformity.

The diversity indices are classified into 5 levels in order of their scope from local to regional, as follows:

- Point diversity - diversity of a single point, a single sampling unit, or a microenvironment;
- Alpha - within the diversity of a habitat; usually made up of several sub-samples from a single habitat; alpha diversity refers to the number and abundance of species in a community;
- Beta - species diversity of a habitat, transect or gradient; beta diversity expresses the difference in species composition and abundance between or within communities;
- Gamma - the diversity of a medium-sized geographical unit, such as an island or micro-watershed; obtained by counting the total number of species within an area (IBGE, 2007);
- Epsilon - the diversity of a broad region, such as a biome, or the catchment area of a large river, of the 5th order or higher.

DIVERSITY INDICES A

Alpha diversity is determined by counting the number of species and the number of individuals of each species, or by accumulated dominance curves for the species sampled (IBGE, 2007).

Species richness

Species richness is only represented by the number of species present in an ecosystem. Comparisons between the species richness of two communities are valid if the sampling intensity in each has been such that the curve of species per area has reached stability at its maximum; otherwise it may be that the true total number of species in the communities has not been obtained to make the comparison.

Relative abundance of species

It represents the percentage of individuals of each species over the total number of individuals. It is the equitability of the species or uniformity of abundance between species.

Berger-Parker Index

The Berger-Parker diversity index is a measure of dominance that expresses the proportional importance of the most abundant species (Newton, 2007), and is defined by the equation:

$$d = \frac{N_{max}}{N}$$

where: d = dominance of the species considered; N_{max} = number of individuals of the most abundant species; N = total number of individuals of all species.

Rényi Entropy Index

Statistical entropy is the measure of the amount of uncertainty in a batch of data.

Alfred Rényi developed an entropy index of order α, generalizing the Shannon entropy index and the Kullback-Leibler divergence index, creating a broad serial diversity index.

Species richness, the Shannon index, the Simpson index, and the Berger-Parker index can be classified as particular cases of Rényi entropy (RÉNYI, 1961, *apud* Bhamidipati, 2009). Rényi entropy of order α, where α ≥ 0, is defined as:

$$H_{\alpha} = \frac{1}{1-\alpha} \, log \sum_{i=1}^{S} \phi_i^{\alpha}$$

Where: $\phi_i = n_i / N$, or fraction of individuals of species i; α = 0, 1, 2, 3,..., ∞.

Shannon-Wiener diversity index

The Shannon-Wiener categorical diversity index is the most widely used (Newton, 2007) and can be calculated by:

$$H' = -\sum_{i=1}^{S} \phi_i \, . \, ln \, \phi_i$$

where: $\phi_i = n_i / N$, or fraction of individuals belonging to species i; n_i = number of individuals sampled for species i; S = number of species sampled; N = total number of individuals sampled from all species.

The Shannon Index assumes that individuals are sampled at random from an infinite population with a random distribution; it assumes that all species in the community are represented in the sample, which is generally not the case and is a source of error in the method (MAGURRAN, 1988). Despite this, it is the most widely used in phytosociology and is recommended for comparison with other forest communities. The higher the H' value, the greater the floristic diversity. The average value is between 1.5 and 3.5; below 1.5 it can be considered low and above 3.5 it is high, and it is rarely higher than 4.5. It is sensitive to the presence of species with low coverage.

Brillouin diversity index

It is an index similar to the Shannon-Wiener index (BARROS, 2007), and is calculated by:

$$HB = (ln\, N! - \sum_{i=1}^{S} ln\, n_i!) / N$$

Where: HB = Brillouin diversity index, n_i = number of individuals of species i; S = number of species; N = total number of individuals of all species.

Jentsch mixing ratio

This index measures the diversity of the population (UBIALLI, 2007) and is calculated by:

$$QM = \frac{S}{N}$$

Where: S = number of species sampled; N = total number of individuals sampled.

The closer the QM value is to 1 (one), the more diverse the population is. The QM value can be presented as a proportion, i.e. by inverting the original N/S expression and presenting the result as a proportion of the number of individuals in relation to the number of species for each plot and for the total.

Pielou's index of equitability

It is an index that measures uniformity between communities (PIELOU, 1977) and is calculated by the expression:

$$C = \frac{H'}{Ln\ (S)}$$

Where: C = Pielou's evenness index; Ln (S) = H_{max} = maximum diversity; S = number of species sampled = richness.

The C value ranges between 0 and 1, where 1 represents maximum diversity, i.e. all species are equally abundant.

The Pielou index of evenness is derived from the Shannon index.

Simpson's index

Edward Hugh Simpson developed a diversity index for ecosystems that indicates the probability that two individuals taken at random in a population belong to the same species (Newton, 2007):

$$D = \Sigma p_i^2$$

Where: D = Simpson's diversity; p_i = proportion of individuals of species i.

For finite communities, the following equation should be used:

$$D = \sum \left(\frac{n_i(n_i - 1)}{N(N - 1)} \right)$$

Where: n_i = number of individuals of species i; N = total number of individuals; $0 \leq D \leq 1$.

As D increases, diversity decreases. Thus, Simpson's index is usually expressed as 1/D. It is sensitive to the presence of species with high coverage.

DIVERSITY INDICES Β

Measures of *β-diversity,* or measures of differential diversity, describe how distinct (or similar) communities are in terms of species composition. When comparing two different communities, the *β* diversity is higher the lower the number of species that occur in both. To obtain it, it is necessary to ensure sampling sufficiency by constructing the species-area curve and calculating similarity indices between sampling units (IBGE, 2007). Diversity indices *β* between pairs of communities can be qualitative or quantitative.

Examples of β diversity measures are the Jaccard and Sørensen, Morisita-Horn and Whittaker indices, described below, and the Cody, Routledge, Wilson and Shmida measures. The similarity coefficients generate very useful measures for comparing three or more communities, making it possible to see which are the most similar (or dissimilar). The indices are calculated for each pair of communities and the results can be compared.

When comparing a large number of communities or gradients simultaneously, the determination of *β* diversity should be carried out by means of cluster analysis.

3.10.2 SIMILARITY BETWEEN COMMUNITIES

Similarity indices are used to compare the flora of two or more plant communities, indicating the degree of similarity between them. The concept of similarity between plant communities was developed by various researchers, with different subjective interpretations, until it began to be expressed mathematically as a function of the number of common species between two communities, or in reverse, expressing dissimilarity by the number of different species that occur in the first and do not occur in the second community. Two groups of similarity indices can be distinguished: taxonomic and biocenotic. The former consider only the presence or absence of each species in each community, while the biocenotic indices take into account the number of individuals of each species present in each biocenosis.

Among the best known and most widely used taxonomic similarity indices are Jaccard and Sørensen. Among the biocenotic indices are Czekanowski and Morisita-Horn.

JACCARD SIMILARITY INDEX

Jaccard's similarity index is simple and is expressed only as the percentage of species common to two communities in relation to the sum of species present independently in both (Mueller-Dombois and Ellenberg, 1974). Jaccard's community coefficient is calculated using the equation:

$$IS_j = 100 \,.\, c / (a + b + c)$$

Where: a = number of species present in community A and absent in community B; b = number of species present in community B and absent in community A; c = number of species common to both communities.

SØRENSEN'S SIMILARITY INDEX

Among the many similarity indices derived from or inspired by Jaccard's, the best known is probably Sørensen's. It is a qualitative index based on the presence or absence of species (Mueller-Dombois and Ellenberg, 1974). It is a qualitative index based on the presence or absence of species (Mueller-Dombois and Ellenberg, 1974), and is calculated by:

$$IS_s = c / (1/2) \,.\, (A+B)$$
$$\text{or, } IS_s = 2 \,.\, c / (A+B)$$

Where: SIs = Sørensen's similarity index; A = total number of species in community A; B = total number of species in community B; c = number of species common to both communities.

The index can be expressed as dissimilarity, or Coefficient of Distance (CD_s), by the following equation:

$$CD_s = 100 - IS_S = 100 - \{ 200.\ c / (A+B) \}$$

CZEKANOWSKI SIMILARITY INDEX

The Czekanowski index is similar to the Jaccard index, but uses qualitative and quantitative data, ranging from 0 (absolute dissimilarity) to 1 (absolute similarity) (IBGE, 2007).

The Czekanowski index (HENRIQUES *et al.*, 1996) is calculated by:

$$IS_c = 2 \,.\, min\ (x_{i,\,i}) / (x + i)_i$$

Where: min ($x_{i,i}$) = lowest number of individuals for the species found in both communities; x_i e i = number of individuals of species "i" found in both areas.

$$Ps_{1,2} = \frac{2\sum_{i=1}^{N} min(X_{i1}, X_{i2})}{\sum_{i=1}^{N}(X_{i1} + X_{i2})}$$

Where: X_{i1} and $Xi2$ = number of individuals of species i in communities 1 and 2, respectively; $min\ (X_{i1}, X_{i2})$ = minimum amount of species i common to both communities.

Or, considering the proportions of the number of species:

$$Ps'_{1,2} = \sum_{i=1}^{N} min(Y_{i1}, Y_{i2})$$

Where: $Y_{i1} = X_{i1}/\sum_{i=1}^{N} X_{i1}$; $Y_{i2} = X_{i2}/\sum_{i=1}^{N} X_{i2}$.

MORISITA-HORN INDEX

This index is little influenced by the size of the sampling unit and the richness of species, and is a good approximation of the degree of dispersion (BARROS, 2007; HENDERSON, 2003, *apud* Newton, 2007). The Morisita-Horn index is more complex than the Jaccard and Sørensen indices and takes into account the abundance of species:

$$C_H = \frac{2.\sum X_{ij}.X_{ik}}{\left[\left(\frac{\sum X_{ij}^2}{N_j^2}\right) + \left(\frac{\sum X_{ik}^2}{N_k^2}\right)\right].N_j.N_k}$$

Where: C_H = Morisita-Horn index; X_{ij} = number of individuals of species i in community j; X_{ik} = number of individuals of species i in community k; N_j = total number of individuals in community j; N_k = total number of individuals in community k.

The percentage of similarity is calculated by:

$$P = 100 - 0{,}5 \sum_{i=1}^{S} (P_{ai} - P_{bi})$$

Where: P = percentage of similarity; P_{ai} = percentage abundance of species i in community a; P_{bi} = percentage abundance of species i in community b; S = total number of species.

The similarity coefficient ranges from 0, when there is not a single common species between the communities, to 1, when all the species that occur in community 1 are present in community 2 and vice versa. Similarity tends to decrease when species abundance is taken into account, which is the case with this index.

WHITTAKER MEASURE

The Whittaker measure (Newton, 2007), like the similarity coefficients, aims to assess how much the species composition changes along a gradient or transect, and is obtained from the equation:

$$\beta w = \left(\frac{S}{\alpha}\right) - 1$$

Where: βW = Whittaker's measure; S = total number of species; α = average number of species per sampling unit.

DIVERSITY BY GROUPING

Cluster analysis is carried out using a matrix of similarity (or dissimilarity) between pairs of samples (Barros, 2007). It is usually carried out to group similar sample units or communities with the intention of defining the different associations present in the survey area. The units with the greatest similarity are combined to form groups, giving rise to a dendrogram. An example of a phytosociological survey is shown in Tabela 9 with five sampling units. Jaccard's index was used to determine the dissimilarity coefficient, the matrix of which is shown in Tabela 10 . In Figura 23 shows the resulting dendrogram. The groups are formed from the units with the least dissimilarity.

Tabela 9 - Matrix of occurrence (1) or absence (0) of species per sampling unit (Sites).

LOCA TIONS	ESPECIES											
	A	B	C	D	E	F	G	H	I	J	K	L
Location 1	1	1	0	1	1	1	1	0	1	0	1	1
Site 2	1	0	1	0	0	0	1	1	0	1	1	0
Site 3	0	1	1	1	0	0	1	0	1	0	1	0
Site 4	1	0	1	1	1	1	0	1	0	1	0	0
Location 5	0	1	0	1	0	0	1	1	1	0	1	1

Tabela 10 - Jaccard dissimilarity matrix (or Jaccard distances)

	Local1	Local2	Local3	Local4	Local5
Local1	0.00000	.	.	.	.
Local2	0.75000	0.00000	.	.	.
Local3	0.50000	0.66667	0.00000	.	.
Local4	0.66667	0.55556	0.81818	0.00000	.
Local5	0.40000	0.70000	0.37500	0.83333	0

The SAS program for analysis is shown below:

```
DATE A;
   LOCAL INPUT $6. SP1 SP2 SP3 SP4 SP5 SP6 SP7 SP8
SP9 SP10 SP11 SP12;
CARDS;
LOCATION1 1 1 0 1 1 1 0 1 0 1 0 1 1
LOCAL2 1 0 1 0 0 0 1 1 0 1 1 0
LOCAL3 0 1 1 1 0 0 1 0 1 0 1 0
LOCAL4 1 0 1 1 1 1 0 1 0 1 0 1 0 0
```

```
LOCAL5 0 1 0 1 0 0 1 1 1 0 1 1
;
PROC DISTANCE DATA=A METHOD=DJACCARD
ABSENT=0 OUT=DISTJACC;
  VAR ANOMINAL(SP1--SP12);
  LOCAL ID;
PROC PRINT DATA=DISTJACC;
  TITLE1 "JACCARD DISTANCE MATRIX";
  LOCAL ID;
GOPTIONS VSIZE=6IN HTEXT=3PCT HTITLE=3PCT;
AXIS1 ORDER=(0 TO 1 BY 0.1);
PROC CLUSTER DATA=DISTJACC METHOD=WARD
SIMPLE TREE=TREE PRINT=14;
  TITLE1 'GROUPING UNITS BY WARD METHOD';
  TITLE2 'AS A FUNCTION OF JACCARD DISTANCE';
  LOCAL ID;
  VAR LOCAL1-LOCAL5;
PROC TREE DATA=TREE NCLUSTERS=14 GRAPHICS
HAXIS=AXIS1;
RUN;
QUIT;
```

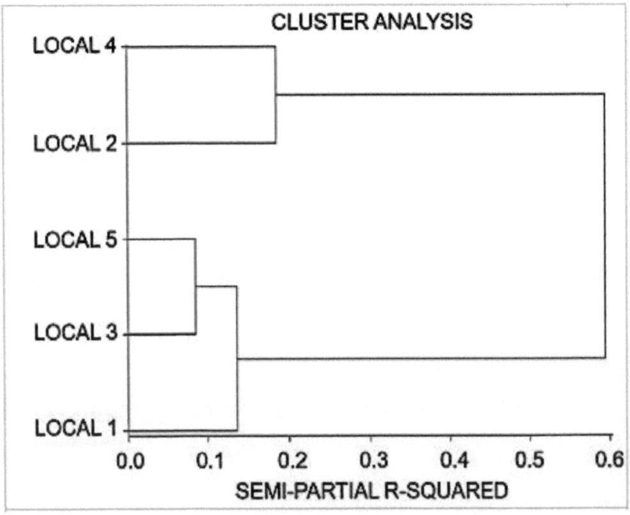

Figura 23 - Grouping of units (Sites) by Ward's method, based on the Jaccard distance matrix.

Among other measures used for cluster analysis is the Euclidean Distance (ED) (GARCIA and LOBO-FARIA, 2007), considered a quantitative index, calculated by:

$$DE = \sqrt{(x^2+y^2)}$$

Where: DE = Euclidean distance; x = difference between the abundance of species A and species B in community or sampling unit 1; y = difference between the abundance of species A and species B in community or sampling unit 2.

SPECIES AGGREGATION

Aggregation indices estimate the spatial distribution pattern of individuals of each species in a community (Figura 24). These include the MacGuinnes, Fracker and Brischle and Payandeh indices.

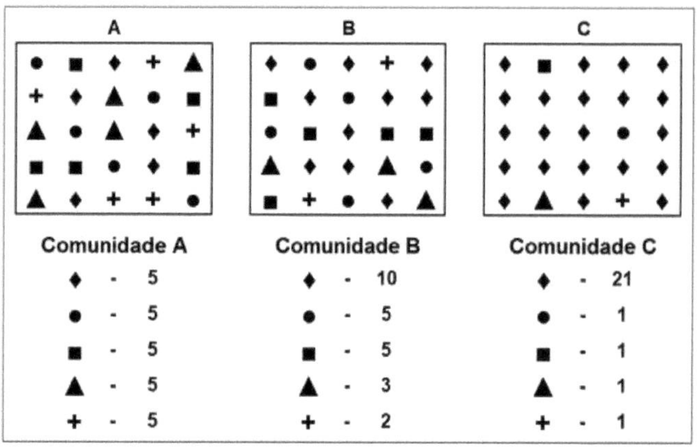

Figura 24 - Horizontal distribution patterns of species in different communities. Source: IB-USP (2009).

MACGUINNES AGGREGATION INDEX

This index estimates the degree of aggregation of the species based on the observed (D_i) and expected (d_i) densities, and is calculated by:

$$IGA_i = D_i/d_i$$

Where: IGA_i = aggregation index of species i; $D_i = n_i / u_t$ = observed density of species i; $d_i = - \ln (1\text{-fi})$ = expected density of species i; $f_i = u_i / u_t$ = observed frequency of species i; n_i = number of individuals of species i; u_i = number of sampling units in which species i occurs; ut = total number of sampling units.

The distribution of individuals of the species is classified as:

- IGAi < 1 - uniform distribution;
- IGAi = 1 - random distribution;
- 1< IGAi≤ 2 - distribution with a tendency to cluster;
- IGAi > 2 - aggregated or grouped distribution.

Fracker and Brischle Index

It estimates the shape of the spatial distribution of species (UBIALLI, 2007) and is expressed by the equation:

$$K_i = (D_i - d_i)/d_i^2$$

Where: K_i = Fracker and Brischle aggregation index for species i; D_i = observed density of species i; d_i = expected density for species i.

The spatial distribution pattern of the species is classified according to the following scale:

- Ki≤ 0.15 - random distribution;
- 0.15 < Ki≤ 1.0 - tendency to cluster;
- Ki > 1 - aggregated or grouped distribution.

Payandeh Index

By estimating this index, the degree of aggregation of the species is obtained through the relationship between the variance in the number of trees per plot (PAYANDEH, 1970) and the average number of trees:

$$P_i = \frac{S_i^2}{\overline{n_i}}$$

$$S_i^2 = \frac{\sum_{i=1}^{n} n_i^2 - \frac{(\sum_{i=1}^{n} n_i)^2}{n}}{n-1} \quad e \quad \overline{n_i} = \frac{\sum_{i=1}^{n} n_i}{n}$$

Where: P_i = Payandeh Index for species i; S_i^2 = variance of the number of trees of species i; n_i = number of trees of species i in plot j; $\overline{n_i}$= average number of trees of species i; n = total number of trees of species i.

The spatial distribution pattern of the species is classified according to the following scale:

- Pi < 1 - random distribution or non-grouping;
- 1≤ Pi≤ 1.5 - tendency to cluster;
- Pi > 1.5 - aggregated or grouped distribution.

LOYALTY INDEX

A faithful species is one that only occurs in a certain type of community. Fidelity can therefore be defined as the level of restriction of a species' occurrence to a given community. Species that occur in a given community and do not occur or occur very little in others are called characteristic species, and are considered faithful to that community. Just as there are plants that live restricted to certain soils or climates, there are also those that are strictly confined to certain communities. Species with absolute loyalty are called exclusive.

It is then possible to define groups of plants characterized by the fidelity of characteristic species by the fact that these species are only present in one or more groups (where their degree of presence is not necessarily great) and absent in all the others.

In order for the fidelity values to be accurate, it is important to recognize and classify the community. Only areas of fully developed vegetation should be used. Areas that have been burned, flooded or subjected to recent human intervention show changes in vegetation, resulting in ecologically unstable associations that are not representative of the original community, and classification using statistical fidelity should be avoided.

The Zurich-Montpellier School recognizes five degrees of loyalty:

- Exclusive species - appear restricted to certain communities, in a particular geographical region;
- Selective species - found especially in one community and may occasionally occur in another;
- Preferred species - present in several communities, but predominant in one of them;
- Indifferent or companion species - have no particular affinity for one type of community;
- Strange or accidental species - are rare in a particular community.

Fidelity is determined by the presence of each species in the sampling units representing the different communities.

According to Matteucci (1982), Goodall proposed a technique for quantitatively defining fidelity. The degree of fidelity is defined as the difference in the frequencies of the species in both communities expressed as a proportion of the lowest frequency:

$$F = \frac{\frac{a}{a+c} - \frac{b}{b-d}}{\frac{b}{b+d}} = \frac{a(b+d)}{b(a+c)} - 1$$

Where: a,b,c,d = sample units from two communities A and B to be compared; a=species x present in A; b=species x present in B; c=species x absent in A; and d=species x absent in B.

When F=0 the species is present (in equal frequency) in both communities; when F tends to infinity the species is confined to community A, i.e. it is faithful to A.

According to Braun-Blanquet, species that are not limited to certain areas are much more abundant, as they are not sociologically restricted to certain plant communities. The close social confinement of characteristic species is the result of causes such as:

- beneficial mutations that have made it possible to increase the vitality of certain populations representing some species separated by geographical isolation or by the formation of small breeding communities, causing these populations to become faithful to a community, and these populations can specialize to the point of forming new endemic species;
- geographic isolation, allowing for persistent inbreeding between species, causing them to lose their capacity for variation and expansion (dissemination) due to closer biotic and abiotic relationships in the area (adaptations to the physical and chemical conditions of the habitat, parasitism, competition, etc;)
- some species may be generalists in certain places, but have exclusive characteristics in other regions, because only in these areas can they satisfy their vital needs;
- lack of similar communities between the major phytogeographic zones;
- some species that could grow well in certain areas don't, because there are better adapted species (competition).

Index of equitability

This index is obtained from the Shannon index. It translates the relationship between actual diversity and the theoretical maximum diversity. It is calculated using the equation:

$$J = H' / H' max$$

Where: J = evenness index; H' = Shannon index= $-\sum_{i=1}^{S} \pi_i \ . \ ln\, \pi_i$; H' max = maximum value of the Shannon index; π_i = n_i /N, or fraction of individuals belonging to species i; n_i = number of individuals sampled for species i; S = number of species sampled; N = total number of individuals sampled from all species.

The value of the index varies between 0 and 1. The closer the equitability is to 1, the more balanced the distribution of species in the community.

Values close to 0 mean that all or almost all individuals are of the same species.

3.10.3 Horizontal structure

The main objective of studying the horizontal structure of a plant community is to determine the phytosociological importance of each species based on the variables density, frequency and dominance, which are used to calculate the coverage index of each species and its importance value in the community.

Density

The total density of a given community is the expression of the total number of individuals in it in relation to the total area considered (UBIALLI, 2007):

$$DT = 10000 \,.\, N / A$$

Where: DT = total density of the community; N = total number of individuals in the community; A = total area of the community in m².

The density of a species in a community is nothing more than the number of individuals per unit area:

$$DA_s = 10000 \,.\, n_s / A$$

Where: DA_s = absolute density of species s; n_s = number of individuals of species s; A = area considered in m².

Species can be classified according to density, in levels of abundance, as follows:

- very rare species - <1%;
- rare species - from 1% to <5%;
- occasional species - from 5% to <25%;
- abundant species - from 25% to <75%;
- very abundant species - 75% or more.

Density, when expressed as a percentage, makes up the species' relative density index, and is calculated using the expression:

$$DR_s = 100 \,.\, DA_s / DT$$

Where: DR_s = relative density of species s; DA_s = absolute density of species s; DT = total community density.

Frequency

Assessing the occurrence of a species throughout a community is done by sampling, locating sampling units systematically or randomly over its area. The frequency with which a given species occurs in the various sampling units is an indicator of the uniformity of its spatial distribution in the community under study. If the species is homogeneously dispersed, it will tend to appear in a large number of sampling units. The more heterogeneous its distribution, the smaller the number of sampling units in which it will appear, and vice versa. The absolute frequency of a species in a community is simply the number of sampling units in which it was found (UBIALLI, 2007), as follows:

$$FA_s = U_s$$

Where: FA_s = absolute frequency of species s; U_s = number of sampling units where species s is present.

There is not much point in this index, as the smaller the sampling units and the greater the sampling intensity, the higher the value found. Therefore, it should always be expressed as a percentage of sampling units in which the species occurs in relation to the total number of units that make up the sample, and can be calculated using the following equation:

$$FR_s = FA_s / U$$

Where: FR_i = relative frequency of species s; FA_s = absolute frequency of species s (or number of sampling units in which species s occurs); U = total number of sampling units.

In Figura 25 there are three communities where species "A" occurs.

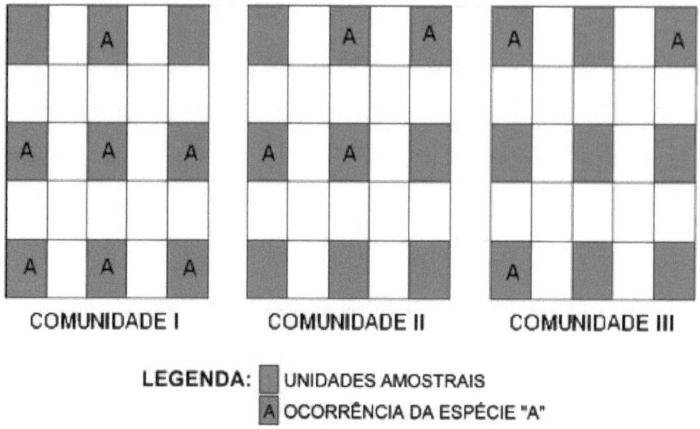

Figura 25 - Occurrence of species "A" in the sampling units of 3 communities of the same size.

In community I the distribution is more uniform than in communities II and III, where the absolute frequencies are 7, 4 and 3 respectively and the corresponding relative frequency indices are 77.8%, 44.4% and 33.3%.

Braun-Blanquet (1979) created the following frequency table:

- 5 or VI - almost always present (between 80-100%);
- 4 0u V - present most of the time (60-80%);
- 3 or IV - often present (40-60%);
- 2 or III - infrequent (20-40%);
- 1 or II - rare (2-20%);
- ss or I - sporadic (less than 2%).

Dominance

Dominance expresses the occupation of space by a given species. Due to the high correlation that the cross-sectional area of the trunk at breast height has with the total occupation of space by a tree, it was agreed to use this variable to represent the dominance of space. The sum of the cross-sectional areas of all the trees of a species per unit area represents the absolute dominance (DoA) of one species over the others in a given community (UBIALLI, 2007).

The percentage of a species' absolute dominance over the total basal area is its relative dominance (DoR). The following equations are used to calculate this:

$$DoA = \frac{\sum_{s=1}^{S}\sum_{i=1}^{n_j} g_{i,s}}{\sum_{p=1}^{P} Area_p} = G$$

Where: DoA = G = absolute dominance of all trees of all species in m²/ha, or basal area per unit area in m²/ha; $g_{i,s}$ = basal area of tree i of species s; S = total number of species; n_s = number of trees of species s; $Area_p$ = surface area of sampling unit p in hectares (ha); P = total number of sampling units.

$$DoA_s = \frac{\sum_{i=1}^{n} g_i}{Área_p}$$

Where: DoA_s = absolute dominance of species s in m²/ha in sampling unit p; g_i = basal area of tree i in m²/ha; Area = surface area of sampling unit p in hectares (ha).

$$DoR_s = \frac{100.\,DoA_s}{DoA}$$

Where: DoR_s = relative dominance of species s; DoA_s = absolute dominance of species s in m²/ha; DoA = absolute dominance of all species in m²/ha.

Coverage Value

Cover is represented by the area occupied by individuals in relation to the total area considered, expressed as a percentage. It can be measured or estimated. According to Dansereau (1946), cover can be determined in various ways, including density, volume, basal area and number of individuals. The same author recommends that it should be estimated independently for each synusia. Cover is expressed as a percentage and the total of a synusia rarely reaches 100% cover.

To estimate the value of a species' cover index, the density and relative dominance of the species in a plant community are added together. This value is considered by some phytosociologists to be the most important for a species as it expresses its occurrence and dominance in relation to the others (UBIALLI, 2007). The cover value therefore varies between 0 and 200% and is calculated by:

$$VC_s = DR_s + DoR_s$$

where: VC = cover value of species s; DR = relative density of species s; DoR = relative dominance of species s.

It is considered a measure of abundance when it is not possible to determine density (in the case of grasses and stoloniferous species). Braun-Blanquet (1979) developed a six-level coverage scale, as follows:

- 5 = continuous or dominant - more than 75% coverage;
- 4 = frequent or interrupted - 50 to 75% coverage;
- 3 = dispersed or common - 25 to 50%;
- 2 = rare - 11 to 25%;
- 1 = very rare - 1 to 10%;
- + = sporadic - less than 1%.

When the vegetation is structured in vertical strata, it is necessary to determine the coverage per stratum. This methodology is recommended for sampling units of up to 10 m^2; in larger units where the diversity is great, it is difficult to estimate the coverage value, except with the aid of large-scale aerial photographs. It should be noted that in seasonal vegetation, cover can vary from one time of year to the next.

When the inventory is carried out on fixed area plots, the direct method for calculating cover can be used, by measuring the surface area of the tree canopies, or indirectly, by estimating the percentage of cover as a function of points distributed randomly or systematically over the community under study and determining the occurrence or not of each species at each point. The latter method is especially useful when it is not possible to identify individuals separately, as in the case of stoloniferous species and grasses when surveying herbaceous vegetation.

ABUNDANCE INDEX

The abundance value is estimated for each species in each sampling unit (DANSEREAU, 1946), and is classified on a scale of Tabela 11 from the most abundant to the least abundant.

Tabela 11 - Scale of abundance

Level	Abundance	Dominance	Coverage
5	Very abundant	Dominant at least in its synusia - pure formation or almost	Highest coverage value 81 to 100%
4	Abundant	Dominant or subdominant	High coverage 61 to 80%
3	Common	Common	Remarkable coverage 41 to 60%
2	Frequent		Restricted coverage 21 to 40%
1	Occasional		Reduced coverage 1 to 20%
+	Rare		Insignificant coverage 0.1%
-	Present		Rare individuals <0,1%

Importance Value

The ecological importance value of a species in a plant community is calculated by adding its cover value and its frequency, as follows (UBIALLI, 2007):

$$VI_s = VC_s + FR_s$$

Where: VI_s = ecological importance value of species s; VC_s = cover value of species s; FR_s = frequency of species s.

3.10.4 VERTICAL STRUCTURE

Forest species can form different horizontal arrangements that result in multiple vertical arrangements, depending on the environment and its evolution over time. Vegetation tends to have the same structure, both horizontal and vertical, for the same succession of edaphoclimatic events, but the evolution of vegetation itself can result in different structures under similar environmental conditions. There are randomizing factors that influence the formation of vegetation such as fauna, wind and rain, and on the other hand there are limiting factors such as microclimate, topography, soil (fertility, humidity and structure) and the characteristics of the species themselves.

In addition to the fact that each species is different in size, occupying different strata of the vegetation (Figura 26), in certain situations a species can have its growth facilitated. While in others, it may be hindered and although present in a large area, there will be parts where it will be more frequent and dense due to more intense regeneration, with more vigorous growth and higher size, while in others, it will be less dense and smaller, suffering more from competition from other species. This means that not only does the horizontal structure vary from site to site, but also the vertical structure. Therefore, the study and characterization of vegetation implies a description of its vertical structure, as well as its horizontal structure, in order to understand it perfectly.

A species' occupation of vertical space is defined as its sociological position in the plant community. By combining the sociological position of species with the characteristics of their regeneration, we have a complete study of the vertical structure of a plant community.

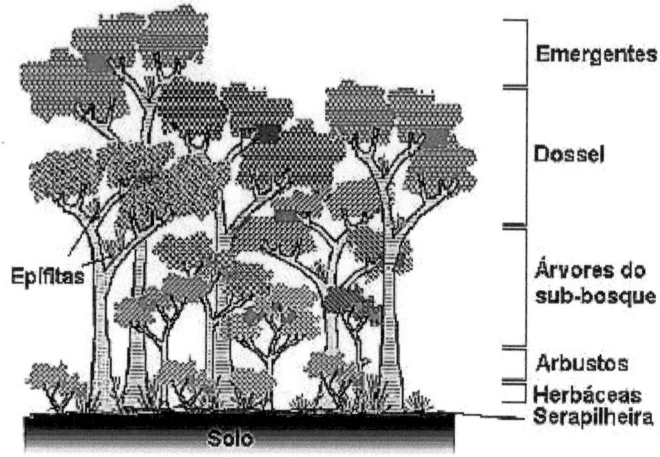

Figura 26 - Vertical structure of a hypothetical forest community. Source: IB-USP (2009).

Sociological position

The different species at different life stages present in a plant community form characteristic vertical strata resulting from the species composition in a given environment, representing their sociological vertical structure or vertical expansion of the forest species. In a natural forest community, the sociological position of a tree is determined by its vertical expansion in relation to its neighbors, which can be described in three arboreal strata, according to Schneider (2002):

- Upper tree stratum - includes the trees whose tops form the highest canopy of the forest; it is made up of the emergent canopies;

- Middle tree stratum - corresponds to trees whose tops are below the highest canopy, but in the upper half of the space occupied by the vegetation; it is the main canopy;
- Lower tree stratum - includes the trees whose crowns are in the lower half of the space occupied by the forest; it is made up of the trees and shrubs of the understory.

Sometimes it is divided into 4 strata, adding a fourth stratum below the lower one, which can be one of the two options below, for native (understorey) or planted (dominated) forests, as follows:

- Understory - shrubs and small trees below the lower stratum; used with native forests in climax or advanced stages of succession;
- Dominated or suppressed - stratum below the lower one made up of small, diseased or genetically inferior trees that were unable to out-compete the others; this type is most commonly used with planted forests.

Mathematically, the three empirical classes are separated by limits referring to three diameter classes, as the height of the tree canopy is difficult to measure in natural forests.

It has been accepted that there are stable and unstable species in a plant community. Stable species are those present in all vertical strata with a diameter distribution forming a negative exponential curve. Species considered unstable are those with an irregular distribution in the different vertical strata. They are more unstable the more irregular their distribution and the less they regenerate, as they tend not to remain in the composition of the plant community.

Natural Regeneration

The natural regeneration of a forest community can be understood as all the seedlings of tree species that have not yet reached tree size. This is a limit that should be set according to the size of the vegetation, but in general, all trees with a DBH of less than 5 cm can be considered to be regenerating. It should be borne in mind that plants up to 1.3 m in height do not have a DBH and should be measured separately, creating two classes of regeneration: plants that can have their DBH measured and those that cannot. Thus, for practical purposes, the existence of at least two classes has been considered, as follows:

- Regeneration 1 - tree and shrub seedlings up to 1.5 m high;
- Regeneration 2 - seedlings of trees and shrubs over 1.5 m tall and less than 5 cm DBH.

There are situations in which the number of small saplings is very large and there is a need to divide Regeneration class 1 into two (Regeneration 1A and 1B, up to 0.75 m in height and up to 1.5 m in height, respectively, for example) in order to facilitate sampling, carrying out it in different sizes of sampling units for each regeneration subclass (1.0 m^2 and 4.0 m^2, respectively, for example). In this case, experience has shown that the sampling units should be large enough to contain more than 20 measurable individuals so that the distribution of regeneration can be well characterized.

Assessing the regeneration of a community can be done using density and frequency, in the same way as for adult individuals. In addition, it is convenient to determine the rate of transfer or entry of individuals from the smallest to the largest classes, using a distribution of height classes with equal intervals. An example of how to determine the rate of entry is given in Tabela 12 where the ratio between the frequency of trees between the largest and smallest size classes represents the rate of plants transferring from a smaller to a larger class (q), calculated as follows:

$$q = \frac{\sum_{c=1}^{C} \sum_{s=1}^{S} n_{s,c} / n_{s,(c-1)}}{C}$$

Where: q = transfer rate from the smallest to the largest classes (average ratio between the frequency of trees in the largest and smallest height classes); C = number of height classes; c = order number of the height class; S = number of species; s = order number of the species; n = number of trees; ns,c = number of trees of species s in class c; ns,(c-1) = number of trees of species s in class c-1.

The "q" index is higher the more the smaller classes enter the larger ones and can be expressed as a percentage. The height class interval must be constant, as in the example in Tabela 12 so that one community can be compared with another.

In communities in the initial colonization phase, there is intense germination of seeds from pioneer species, but few survive and, of those that do, few manage to transfer to the larger size classes; thus, a low q value is expected. On the other hand, in climax communities, there is little seed germination on the forest floor and a large proportion of the germinated seeds manage to form seedlings that grow and reach the larger size classes, resulting in a higher transfer rate.

Tabela 12 - Transfer ratio of trees from the smallest to the largest classes (hypothetical data).

Species	Class (c)	Height (m)	$n_{s,c}$	$q_{s,c}$	q_c
A	1	< 0,75	25	0,50	0,59
B			20	0,67	
C			40	0,80	
D			30	0,38	
A	2	≥ 0,75 e < 1,50	10	0,40	0,48
B			10	0,50	
C			20	0,67	
D			10	0,33	
A	3	≥ 1,50 e < 2,25	3	0,30	0,40
B			5	0,50	
C			10	0,50	
D			3	0,30	
A	4	≥ 2,25	1	0,33	0,36
B			3	0,60	
C			5	0,50	
D			0	0,00	
Average			n_s	q_s	q
A			39	0,38	0,45
B			38	0,57	
C			75	0,62	
D			43	0,25	
Total			195		

3.11 Growth

The aim of studying tree growth is to learn about the different factors in the environment (exogenous) and in the trees themselves (endogenous) that influence it, as well as the effect of the different treatments they are subjected to during their lifetime.

3.11.1 Factors affecting growth

The exogenous factors that influence growth are environmental factors such as competition between plants, soil, climate, altitude, latitude and topography; while the endogenous factors relate to the genetic potential, size and age of the plant. All these factors will influence the physiology of plants and their ability to grow.

According to Kramer and Kozlowski (1972) the most important physiological processes and conditions related to growth are:

- Photosynthesis: the synthesis of carbohydrates from carbon dioxide and water carried out by trees in the presence of light provides the basic food for other processes;
- Nitrogen metabolism: the incorporation of inorganic nitrogen into organic compounds allows for the subsequent synthesis of proteins and cell protoplasm;
- Fat metabolism: synthesis of fats and related compounds;
- Respiration: the oxidation promoted by respiration, which takes place in the absence of light, releases the energy from food used in assimilation, mineral absorption and other energy-consuming processes;

- Assimilation: this is a fundamental process in growth, responsible for converting food into new protoplasm, cell walls and other substances;
- Food storage: the storage of starch and other nutritious substances in the cells of seeds and woody structures guarantees a food supply for germination and other processes and in times of scarcity;
- Salt accumulation: can be harmful at high levels, but moderate levels of salt ensure optimum growth, related to the cells' ability to accumulate ions in the vacuole, reducing the toxic amount in the cytosol;
- Digestion: converting complex or insoluble forms of food (such as starch) into simpler and more soluble ones that can be transported and used in respiration and other processes;
- Absorption: this is represented by the entry of water, minerals from the soil and oxygen and carbon dioxide from the air;
- Transportation: the movement of water, air, nutrients and hormones within trees.

Kramer and Kozlowski (1972) also list the most important environmental factors at each stage of growth, separating the life of trees into 5 stages (Tabela 13): 1) seed germination; 2) seedling establishment; 3) vegetative growth; 4) reproduction; and, 5) decrepitude.

Trees grow by adding cells. The seed embryo contains primary meristems at the tips of the radicle and the caulicle; both are activated by gibberellins which promote primary growth, also known as elongation. When the primary meristem starts producing cells, it also starts the process of cell differentiation which enables the formation of secondary meristems, activated by auxins and which promote secondary growth, or thickening.

Tree growth is complex and biometric variables can be described mathematically to characterize stands and their productive potential. Growth in different genotypes and environments allows species to be selected and improved for each environment. Other stages of management, such as planning pruning and thinning, depend on knowledge of growth.

Tabela 13 - Important environmental factors per tree life stage

Life stage of trees	Processes and conditions	Most important environmental factors
Seed germination	Water absorption Digestion Respiration Assimilation	Temperature Water Oxygen
Seedling establishment	Photosynthesis Assimilation Transport Internal water balance	Light Water Temperature Nutrients
Vegetative growth	Photosynthesis Respiration Assimilation Transport Internal water balance	Light Water Temperature Nutrients
Reproduction	Photosynthesis C/N balance Flower predisposition Initiation of flower primordia Food accumulation	Light Nutrients Temperature
Decrepitude	Unknown (probably water and hormonal relationships, transport and balance between photosynthesis and respiration	Water Nutrients Pests and diseases

Source: Kramer and Kozlowski (1972).

The study of tree growth is essential for the proper management of forests in a sustainable manner and serves as a basis for planning all forestry production activities. The factors that influence tree growth are listed in Figura 27 below.

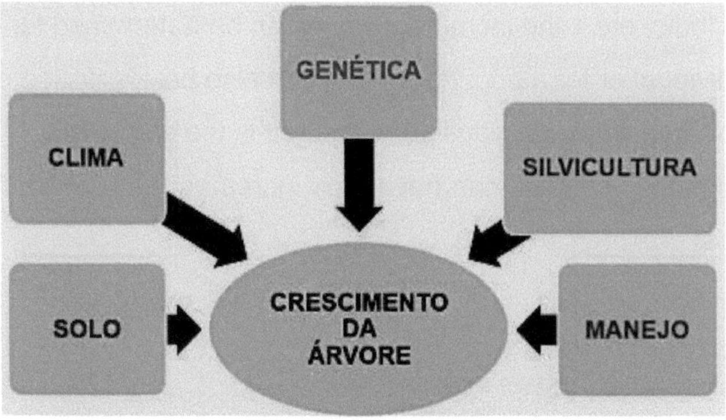

Figura 27 - Factors influencing growth and production.

3.11.2 Mathematical representation of growth

Practically all natural phenomena begin with a slow rate of growth which increases until a certain point, when they reach their maximum; then they start to develop more slowly until growth ceases. Mathematical integrals represent sums of consecutive phenomena and when you try to graph the accumulated growth of a natural phenomenon, such as the growth in height of trees, the curve has a sigmoid shape similar to the integration symbol, which is similar to the integral growth curve shown by most natural phenomena.

3.11.3 Increments

The increment is the growth in the value of the variable over a given time interval (Imanã et al., 2005) and can be calculated mathematically as the first derivative of the growth function.

The increment can be differentiated depending on the time elapsed in which the growth is measured (annual, periodic, etc.); the increment value can be determined for an individual or for a population, and can also be expressed per unit area when the variable measured is the same. All biometric variables can have their increment measured; the most common are: d, g, G, h, h_{dom} , v, and V.

Types of Increments

Annual current increase (ACI)

This is the growth between the start of the growing season and the end of the vegetative resting season over a 12-month period, calculated as:

$$ICA = Y_{(t+1)} - Y_{(t)}$$

Where: t = reference time, year or age; Y = dimension considered.

Percentage increase (p)

$$p = 100\ .\ (Y_{(t+1)} - Y_{(t)}\ )\ /\ Y_{(t+1)}$$

Periodic increase (IP)

This is the total growth over a given period of time:

$$IP = Y_{(t+n)} - Y_{(t)}$$

Where: n = time period; when n = 1 year, IP = ICA.

Annual Periodic Increase (IPA)

This is the average growth per year over a period of time, used for slow-growing species:

$$IPA = (\ Y_{(t+n)} - Y_{(t)}\ )\ /\ n$$

Where: Y = dimension considered; t = reference time, year or age; n = time period in years.

Average annual increase (IMA)

This is the average growth per year up to the date in question, calculated by:

$$IMA = Y(t) / t$$

Where: Y(t) = Value of the variable at the age considered; t = age.

3.11.4 Growth curve

The cumulative growth curve usually has a sigmoid shape and the area under the curve is represented by an integral function (Figura 28).

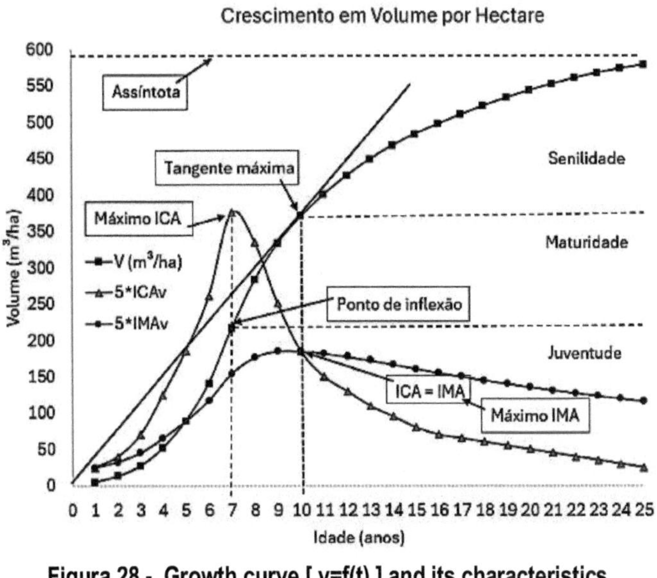

Figura 28 - Growth curve [y=f(t)] and its characteristics.

In youth, growth accelerates and the curve is concave, up to the point of inflection, when annual current growth (ACI) is maximum, the point at which maturity begins; then growth gradually slows down and when average annual growth (ACI) and ACI equalize, the curve forms a maximum tangent with the origin, the point at which ACI is maximum and senility begins. The ICA is the first derivative [f '(t)] of the growth curve. The maximum of the curve is represented by its asymptote. The age (t) at which the IMA is maximum is considered the technical rotation age, as it is the age at which maximum productivity occurs and can be determined by equating the first derivative [ICA=f '(t)] with the average annual increment calculated by [IMA = f(t) / t], resulting in the equation: t = f(t) / f '(t).

3.11.5 Methods for studying growth

The growth (Y) of a biometric variable is a function of time, the environment and the plant. If the environment does not change, growth can be considered a function of time alone: Y = f (t). Therefore, the study of growth must take into account the measurement of the variable in question and the time elapsed or age. Growth is mainly studied for:

- making predictions for the future and planning production;
- evaluate the quality of the environment;
- evaluate the genetic quality of individuals and populations;
- evaluate the quality of the interaction between genotype and environment.

The main methods of measuring variables for studying growth (Imanã Encinas et al., 2005):

- Permanent plots measured annually or periodically;
- Individual trees identified and measured continuously or periodically (tape);
- Trunk analysis
    - full trunk analysis (Figura 31 and 35) - felled trees;
    - partial trunk analysis (Figura 29 and 33) - standing trees (Pressler auger).

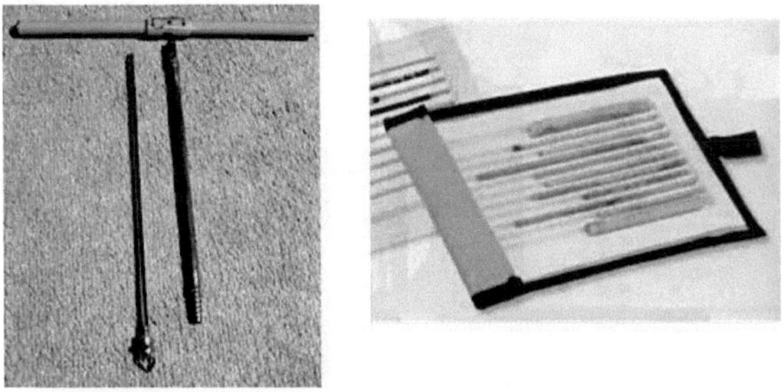

Figura 29 - Partial log analysis (auger and samples).

AGE ESTIMATES IN GROWTH STUDIES

The age of the trees can be estimated by:

- observation and recording of individual trees or permanent plots with individually identified trees whose planting date is known;
- Counting verticils of species with monopodial growth - allows calculation of the IMA_d and estimation of all increases in height;
- Growth ring count.

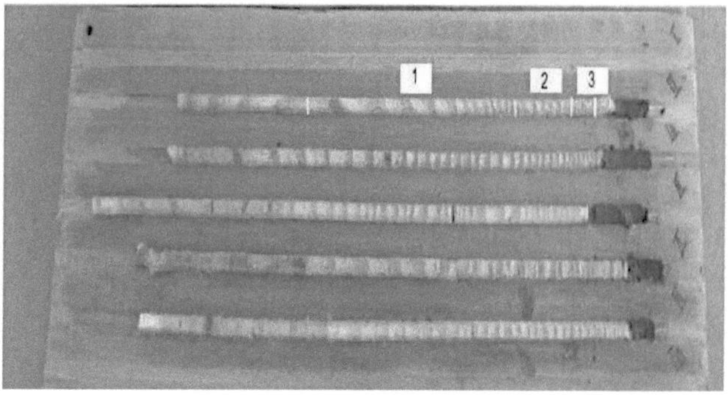

Figura 30 - *Pinus taeda* increment cylinders. Source: Coelho and Hosokawa (2010).

Figura 31 - Complete trunk analysis (twisting and extraction of slices).

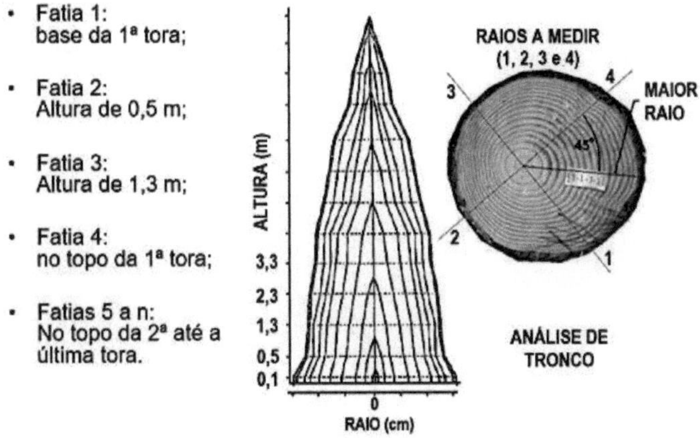

Figura 32 - Complete trunk analysis (twisting).

The growth rings can be measured using a table magnifying glass with a diameter of 9 cm or more and a magnification of 10x or more and a caliper (Figura 33).

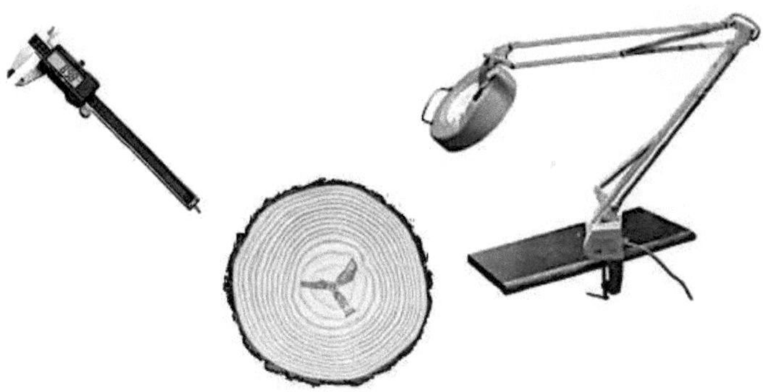

Figura 33 - Measuring growth rings with an articulated magnifying glass and caliper.

Or with devices developed specifically for this purpose, such as the Lintab from the German company Rinntech (Figura 34), which consists of a magnifying glass with a movable table fitted with a micrometer, coupled to a computer with Tsap software which records the measurements for later processing.

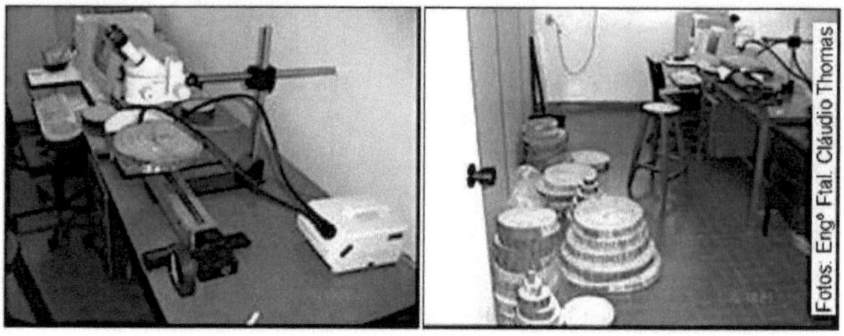

Figura 34 - Measuring growth rings with LINTAB (hardware) + TSAP (software).

4 FORESTRY MODELS IN PRODUCTION PLANNING

A model is a simplified representation of a part of reality. Models can be mental, verbal, material or mathematical. Models can represent objects or systems. Objects are any concrete thing that exists, such as a rock or a screw. Systems are groups of things (composed of matter or energy) that work together, such as the climate, or a forest, or a flow of wood production for certain purposes.

Models have their own characteristics that identify them. The first characteristic is their nature, which can be material or abstract. Material (or physical) models are those produced in the likeness of what they represent, such as a model of a plot of land. An abstract model is a mental, verbal, graphic or mathematical description of something real. A mental model is an idea of something, for example, when someone thinks of a tree, they have an idea of what it looks like, which allows them to recognize it when they see it and distinguish it from grass.

The mental model can be translated into a verbal model, such as the description of a tree: a large plant made up of roots, trunk and crown. The same model can be more detailed, such as: a tree is a plant of the class of gymnosperms, or angiosperms, composed of roots, trunk and crown, whose total height when adult is more than 7 meters and the diameter of the trunk is greater than 5 cm at 1.3 m from the ground. In other words, a model can be simple or detailed.

Models can generate results that are always the same, such as musical scores, which are said to be deterministic; or they can generate results with a certain probability of occurrence, such as experimental designs, which are said to be stochastic, or probabilistic.

Models can be of two types: physical or abstract. Physical models are any physical representation of objects or systems, usually to scale. Examples are models of terrain, buildings, vehicles and airplanes. Abstract models are descriptive, graphic or mathematical representations of objects or systems.

Models can also be classified in terms of time (static or dynamic), type of equation (linear or non-linear), type of prediction (stochastic or deterministic), objectives (analytical or predictive) and variables (discrete or continuous).

Stochastic models are related to a probability of error and are fitted to data from several observations; this is the case with regression equations. Deterministic models result in exact values that can be based on a single observation or individual, for example, tree cubing models such as Smalian, Huber, Newton and Hohenadl.

Developing a model is a process whereby a set of current and historical data is exhaustively analyzed in order to define a rule that will allow a certain future event to be predicted.

In business, predictive models usually exploit patterns found in historical and transactional data to identify risks and opportunities. As such, the models capture relationships between various factors that make it possible to assess the risk or potential associated with a particular set of conditions and thus make better decisions or choices.

Production models are planning tools used to define forestry policies at regional and national level. A production model should correctly predict forest production over time under alternative exploitation hypotheses or subject to different physical and biotic agents (TOMÉ, 2002).

Planning the production of forest stands must take into account three main elements: the technical (growth and production), the economic (financial and social) and the environmental (conservation of biodiversity, soil, water and air).

Time is another important element in forest management and is directly included in the calculations for prognosis as an independent variable. In regulating thinning for Pinus taeda, for example, Schulte et al. (1998) took into account the following parameters relating to time: the date the stand was planted, the date of the first thinning, the length of time between thinnings and the rotation.

Tree growth is mainly influenced by (WYCOFF et al., 1982):

- species - varying with local adaptation and genetic improvement;
- site conditions - represented by slope, aspect, altitude and type of habitat;
- stand conditions - structure and occupation of vertical and horizontal space, regulated by forest management;

- dimensions of the tree itself - represented by DBH, height, crown length ratio and value class.

4.1 Classification of mathematical models

Real models are made up of a series of sub-models of different types and use different modeling methods. Classification aids the modeling task. Models can be, for example:

- Simple or complex;
- Clarification or prediction;
- Empirical or deterministic;
- Static or dynamic.

4.1.1 Classes of modeling objectives

The objectives of modeling can be classified as:

- Structural description - Structure links and retroactive and interactive effects;
- System behavior - mathematically describes the connections and other interactions of the structural system and investigates causes;
- Prognosis - only the system's resulting reactions matter. The internal mechanism is seen as a "black box". Example: canopy projection instead of leaf area; rate of interest.

4.1.2 Theoretical and empirical classification

Empirical mathematical models: with little data, they mathematically describe a theory. Example: $v = f_{1,3} .h.\pi.d^2/4$.

Empirical-statistical models: without much theory, with lots of data, they generate regressions. Example: $v=b +b_{01} .d^2.h$.

4.1.3 Classification by the technique used

Classes are characterized by the resource used to process the data, such as:

- Tables;
- Functions;
- Diagrams;
- Logical relationships (if ... then ...).

4.1.4 Classification by relation to time

Static models: model the state of a system at a given moment; example: $PPL=PPL_{max}\ .(1-e^{-b.t}\ )$, where PPL=net primary production, PPL_{max} =maximum primary production, x=precipitation, b=coefficient.

Dynamic models: these describe how a state changes over time, for example, the height at which the coniferous canopy is inserted changes over time due to natural pruning.

4.1.5 Classification by type of forecast

Deterministic models: these are models that do not allow for random variations; two simulations with the same data produce identical results.

Stochastic models: these are models that can be modified by random numbers (with a normal or uniform distribution) similar to reality. A simulation is a sample. The result is a distribution of values. It can predict the probability of the prediction occurring.

4.1.6 CLASSIFICATION ACCORDING TO DATA TYPE

Models can be classified in terms of their object as models for single trees and models for stands.

4.2 Linear modeling by variable selection

Linear modeling by variable selection is available in SAS, using the STEPWISE, FORWARD, BACKWARD and RSQUARE methods, which are options in the MODEL statement of the software's PROC REG procedure. This makes it possible to carry out various transformations of each independent variable (x) and then select the most significant ones to keep in the model, such as: x, x^2, x^3, x^4 , x^5 , $1/x$, $1/x^2$, $1/x^3$, $1/x^4$, $1/x^5$, $\ln x$, $\ln x^2$, $\ln x^3$, $\ln x^4$, $\ln x^5$, $\ln 1/x$, $\ln 1/x^2$, $\ln 1/x^3$, $\ln 1/x^4$, $\ln 1/x^5$, $x^{1/2}$, $x^{1/3}$, $x^{1/4}$, $x^{1/5}$, among others. Care must be taken not to leave independent variables of lesser significance in the resulting model that produce variance inflation, which can be checked using the VIF (*Variance Inflation Factor*) option in the MODEL statement. The program code could be as follows:

```
FILE DATE;
INPUT X Y;
X2=X**2; X3=X**3; X4=X**4; X5=X**5; IX=1/X;
IX2=1/X2; IX3=1/X3; IX4=1/X4; IX5=1/X5; LNX=LOG(X);
LNX2=LOG(X2); LNX3=LOG(X3); LNX4=LOG(X4);
LNX5=LOG(X5); LNIX=LOG(IX); LNIX2=LOG(IX2);
LNIX3=LOG(IX3); LNIX4=LOG(IX4); LNIX5=LOG(IX5);
RX2=X**(1/2); RX3=X**(1/3); RX4=X**(1/4);
RX5=X**(1/5);
DATALINES;
6 7
8 9
9 10.5
```

```
10 12.5
12 13
14 14
16 15
18 16
20 17.5
22 18.5
24 19
26 20
20 20.5
30 20.5
;
PROC REG DATA=FILE;
MODEL Y = X X2 X3 X4 X5 IX IX2 IX3 IX4 IX5 LNX LNX2
LNX3 LNX4 LNX5 LNIX LNIX2 LNIX3 LNIX4 LNIX5 RX2
RX3 RX4 RX5 / SELECTION=STEPWISE VIF;
RUN;
```

This type of method makes it much easier to find functions to estimate different variables in the forest environment. Unfortunately, there is no analogous procedure for modeling non-linear functions; the latter have to have their models predetermined and adjusted individually.

The Variance *Inflation* Factor (VIF) helps to detect collinearity. It is useful when there are many variables in the model. Some authors suggest that the VIF value should not exceed 3, others suggest 5, but the maximum tolerance value is considered to be 10. When it is too high (>10), the equation can become very distorted, exaggerating the Coefficient of Determination (R^2), greatly reducing the Standard Error of Estimates (Syx), becoming particular to the sample and losing its generality to the population. The IVF can be calculated by forming new models with only the independent variables (X) from the original model. First, the partial R^2 of each independent variable is determined by creating new models with only the independent variables from the original model, making each independent variable from the original model the dependent variable in the new model and all the other independent variables remain independent. The R^2 of each of these new adjusted equations is the partial R^2 of the variable that was made dependent. The IVF is then calculated using the following formula:

$$FIV_j = 1 - (1 / R^2_j)$$

Where: FIV_j = variance inflation factor of variable j; R^2_j = coefficient of partial determination of variable j; j = number of the independent variable in the original model that was made dependent on the other independent variables in the original model.

4.3 Hypsometric relationship models

The hypsometric relationship is the relationship between tree height (h) and diameter (d) and can be represented by the function h = f (d). Müller (2004) tested the 19 models presented in Tabela 14 .

Many models have been created with some transformation of the dependent variable, which should be avoided. If you use a model of this type, such as those listed in Tabela 14 avoid transforming the dependent variable, which is possible by making only the variable to be estimated on the left-hand side of the equation, or by extracting the antilogarithm when it is a logarithmic transformation.

Tabela 14 - Hypsometric relationship models used by Müller (2004).

Nº	Height Function Models
01	$\frac{h}{d} = b_0 + b_1.d$
02	$\frac{h}{d} = b_1.d + b_2.d^2 + b_3.d^3 + b_4.\ln d$
03	$h = b_0 + b_1.d^2$
04	$\ln h = b_0 + b_1.\ln d$
05	$\ln h = b_0 + b_1.d$
06	$\frac{1}{\sqrt{h - 1.3}} = b_0 + b_1.\frac{1}{d}$
07	$h - 1.3 = b_0 + b_1.d + b_2.d^2$
08	$h = b_0 + b_1.d + b_2.d^2$
09	$h = b_0 + b_1.d + b_2.d^2 + b_3.d^3$
10	$h = b_0 + b_1.\frac{1}{d^2}$
11	$\frac{1}{h - 1.3} = b_0 + b_1.\frac{1}{d} + b_2.\frac{1}{d^2}$
12	$\frac{d^2}{\sqrt{h - 1.3}} = b_0 + b_1.d + b_2.d^2$
13	$\ln(h - 1.3) = b_0 + b_1.\frac{1}{d}$
14	$h - 1.3 = b_0 + b_1.d$
15	$\ln h = b_0 + b_1.\frac{1}{d}$
16	$h = b_0 + b_1.d$
17	$\ln(h - 1.3) = b_0 + b_1.\ln d$
18	$\ln(h - 1.3) = b_0 + b_1.\ln d + b_2.\ln^2 d$
19	$\ln(h - 1.3) = b_0 + b_1.\ln\left(\frac{d}{1 + d}\right)$

Where: d = Diameter at 1.30 m height, in centimeters; h = Tree height, in meters; b_0, b_1, b_2, b_3 and b_4 = Equation coefficients; ln = Natural logarithm.

4.4 Models for trunk volume

Many mathematical models have been developed for volumetric estimates (Tabela 15) based on diameter and height (v = f (d, h)). A study with an excessive number of models was carried out by Silvestre et al (2014) who adjusted 33 models for volumetric estimates of *Pinus taeda* L. stands in the municipality of Lages, SC, listed in Appendix A.

Tabela 15 - Examples of mathematical models for trunk volume

Model	Author	Year
$v = b_0 + b_1 . g$	Kopeski-Gerhardt	1899
$v = b_0 + b_1 . d + b_2 . d^2$	Hohenadl-Krenn	1944
$v = b_0 + b_1 . d^2$	Dissescu-Stanescu	1956
$v = b_0 + b_1 . d^2.h$	Spurr	-
$v = b_0 + b_1 . \ln(d^2.h)$	Modified Spurr	-
$v = b_0 + b_1 . d$	Burkhault	-
$v = b_0 + b_1 . \ln(d)$	Burkhault	-
$v = b_0 + b_1 . d + b_2 . h$	Bonnets	-
$v = b_0 + b_1 . d^2 + b_2 . h$	Bonnets	-
$v = b_0 + b_1 . d^{b2} . h^{b3}$	-	-
$v = b_0 + b_1 . d^{b2} + b_3 . h^{b4}$	-	-
$v = b_1 . d^{b2} . h^{b3}$	-	-
$v = b_0 + b_1 . d^2.h + b_2 . d^2$	Ferreira	1993
$v = b_0 . (d^2.h)^{b1}$	Zapeda	1993

Very complex models with many independent variables should be avoided, as the adjustment can make the equation particular to the data from the trees being harvested and no longer for the population as a whole. This can be avoided by calculating the Variance Inflation Factor (VIF) and eliminating variables with a VIF greater than 10, but it is advisable, for greater safety, to avoid variables with a VIF greater than 3. Sometimes the Student's t value calculated for the constant (b_0) in the equation is not significant and should be excluded.

4.5 Form factor models

Although the form factor has fallen out of use due to more appropriate technologies for estimating volume, there are still situations in which it is necessary to study the shape of trees using this variable. Loetsch et al (1973) listed 11 models for estimating the shape factor based on tree diameter and height (Tabela 16).

Tabela 16 - Equation models for form factor.

Number	Models
1	$f = b_0 + b_1 .1/d^2 h + b_2 .1/h + b_3 .1/d^2$
2	$f = b_0 + b_1 .1/d^2 h + b_2 .1/dh + b_3 .1/d + b_4 .1/h + b_5 .1/d^2$
3	$f = b_0 + b_1 .1/h + b_2 .h/d + b_3 .h/d^2$
4	$f = b_0 + b1.d /d_{0,3}{}^2 + b_2 .h/d^2$
5	$f = b_0 + b_1 .d /d_{0,3}{}^{22} + b_2 .h/d^2$
6	$f = b_0 + b_1 .d /d_{0,3}{}^2 + b_2 .h/d^2 + b_3 .1/d$
7	$f = b_0 + b_1 .(d_{0,1} .d_{0,5})/d^2 + b_2 .d /d_{0,3}{}^2$
8	$f = b_0 + b_1 .(d_{0,1} .d_{0,5})/d^2 + b_2 .d /d_{0,3}{}^{22} + b_3 .h/d^2$
9	$f = b_0 + b_1 .(d_{0,5} /d)^2 + b_2 .(d_{0,3} /d)^2$
10	$f = b_0 + b_1 .log(d) + b_2 .log(d_{0,3}) + b_3 .log(h)$
11	$f = b_0 + b_1 .logd + b_2 .logh$

Where: f=shape factor; d = trunk diameter at 1.30 m height; d_0 ,i = Hohenadl diameters in centimeters; h = height in meters. Source: Loetsch et al (1973).

4.6 Shell models

It is very difficult to obtain good equations for estimating tree bark thickness (e) as a function of variables such as diameter (d), height (h), age (h) and diameters relative to height ($d_{0,i}$). Functions of the type e = f (d, h, t, $d_{0,i}$) generally result in a high coefficient of variation and a low coefficient of determination. However, the diameter without bark is easily estimated using the diameter with bark at any position on the trunk and the diameter with bark can be estimated for any position on the trunk using thinning functions. But nature is unpredictable and it is possible that good equations can be found in some cases. To this end, Schneider cites models listed by Loetsch et al (1975) (Tabela 17).

Tabela 17 - Equation models for estimating bark thickness.

Number	Models	Author
1	$e = d / (b_0 + b_1 . d)$	Korsun
2	$e = b_0 + b_. \log(d)$	-
3	$e = b_0 + b_. d + b_2 . t + b_3 . d.t$	Laar
4	$e = b_0 + b\, d_{1.}$	Meyer
5	$e = b_0 + b_{1.} \log(d)$	Modified Mayer
6	$e = b_0 + b_1 . \log(d) + b_2 . \log(d.h) + b_3 . \log(t.h)$	Schneider & Silva modified

Where: e = bark thickness in centimeters; t = age in years; d = diameter at 1.3 m height; h = height in meters. Source: Schneider (2002).

Bark weight is an important variable when you want to manufacture products obtained from the bark, such as tannin. Schneider (2002) lists the 5 models in Tabela 18 :

Tabela 18 - Equations for estimating shell weight.

Number	Models	Author
1	$\log y = b_0 + b_. \log d^2h$	Clark III & Schroeder (1977)
2	$\log y = b_0 + b_1 .\log d$	Kraffenbauer & Andrae (1977)
3	$\log y = b + b_{01} .\log d + b_. \log h$	Meyer
4	$\log y = b_0 + b\ v_.$	Meyer
5	$\log y = b_0 + b_1 .\log d_g$	Loetsch et al. (1975)

Where: y = bark weight; d = diameter at breast height; h = total height; dg = average basal area tree diameter; v = tree volume. Source: Loetsch et al (1975) *apud* Schneider (2002).

4.7 Forest growth models

Forest models can be classified into three main types in terms of the object to be modeled:

- Growth models for singular trees;
- Growth models for size classes (dap);
- Growth models for stands.

Forest growth modeling has the following objectives:

- Understanding growth processes;
- Help with concrete silvicultural and economic options;
- Forecasting development in different scenarios.

Kiviste et al (2002) made an exhaustive compilation of the growth models used in forestry in the book "*Funciones de crecimiento de aplicación en el ámbito forestal*".

GROWTH MODELING

Growth can be modeled using empirical or phenomenological mathematical models, for example:

- Empirical: polynomials and Backman's function;
- Phenomenological: Mitscherlich and Gompertz equations.

Considering growth in years, the integral growth function is represented by:

$$y_t = f\,(t)$$

Its 1st derivative (y') is the annual current increase (ICA):

$$y\,' \approx ICA_t = y_t - y_{t-1}$$

And the average annual increment (IMA) is given by the integral divided by time:

$$IMA_t = y \approx y_t \,/\, t$$

Where: y_t is the value presented by the variable (d, h, v, or other) at age t; y_{t-1} was the value presented by the variable in the previous year t-1; t is the age in years; ICA_t is the increase suffered in the last year (t); IMA_t is the average of the increases up to the current age t.

The Chapman-Richards model (Sit, 1994), also known as the Mitscherlich model, is one of the most widely used in forestry for growth modeling:

$$Y = A\,.\,(1 - e\,)^{-k.tr}$$

Where: A, k, r = are the parameters of the equation; A = is associated with the upper asymptote; k = is associated with the speed of growth (0 to 1); r = is associated with the inflection point of the curve; Y = is the dependent variable (dendrometric variables); t = independent variable (time).

HEIGHT GROWTH

The height growth curve (h) is also known as the height growth integral and can be represented as a function of age:

$$h = f(age), \text{ or } h = f(t)$$

Where: h = height in meters; t = age in years.

The increase in the tree's height each year is known as the Current Annual Increase in Height (ICAh), which is the difference in height from one year to the next:

$$ICAh_{(t)} = h_{(t)} - h_{(t-1)}$$

Where: $ICAh_{(t)}$ = current annual increase in height at age t; $h_{(t)}$ = height at age t; $h_{(t-1)}$ = height at age t-1; t = age.

The average increase up to the age in question is called the Average Annual Increase in Height (AIAh), which is nothing more than the height in the year in question divided by the age of the tree in that year:

$$IMA_{h(t)} = h_{(t)} / t$$

Where: $IMA_{h(t)}$ = average annual height increase at age t; $h_{(t)}$ = height at age t; t = age.

Suppose a tree seedling is planted at a height of 0.30 m and has grown for 7 years until it reaches a height of 21 meters (Figura 35). Consider that it has had annual increments of:

- 2.2 m in the first year, reaching 2.5 m in height;
- 3.5 m in the second year, reaching a height of 6 m;
- 4.0 m in the third year, reaching a height of 10 m;
- 3.5 m in the 4th year, reaching 13.5 m in height;

– 3.0 m in the 5th year, reaching 16.5 m in height;
– 2.5 m in the 6th year, reaching a height of 19 m;
– and 2.0 m in the 7th year, reaching 21 m in height.

Therefore, the tree presented the Annual Current Increment (ICAh), Height (h) and Average Annual Increment (IMAh), listed in Tabela 19 e Figura 36 .

Tabela 19 - Growth of a tree.

Age (t)	ICA_h (m)	h (m)	IMAh (m)
0 = planting	0,3	0,3	0,3
1	2,2	2,5	2,5
2	3,5	6,0	0,6
3	4,0	10,0	0,6
4	3,5	13,5	0,6
5	3,0	16,5	0,6
6	2,5	19,0	0,6
7	2,0	21,0	0,6

Fitting a 5th degree function to the growth integral as:

$$f(t) = b_0 + b_1\, t + b_2\, t^2 + b_3\, t^3 + b_4\, t^4 + b_5\, t^{5}$$

Its first derivative is the formula:

$$f\,'(t) = b_1 + 2\, b_2\, t + 3\, b_3\, t^2 + 4\, b_4\, t^3 + 5\, b_5\, t^4$$

The IMA is given by:

$$IMA = f(t) / t = (f(t) = b_0 + b_1\, t + b_2\, t^2 + b_3\, t^3 + b_4\, t^4 + b_5\, t^5) / t$$

And since the maximum IMA can be determined by equating the first derivative with the IMA [f(t) / t = f '(t)], then the age at which the maximum IMA occurs can be found by dividing the integral function [f(t)] by the first derivative [f'(t)], giving t = f(t) / f '(t), as follows:

$$Idade\ de\ IMA\ máximo = t$$
$$= \frac{b_0 + b_1\, t + b_2\, t^2 + b_3\, t^3 + b_4\, t^4 + b_5\, t^5}{b_1 + 2b_2\, t + 3b_3\, t^2 + 4b_4\, t^3 + 5b_5\, t^4}$$

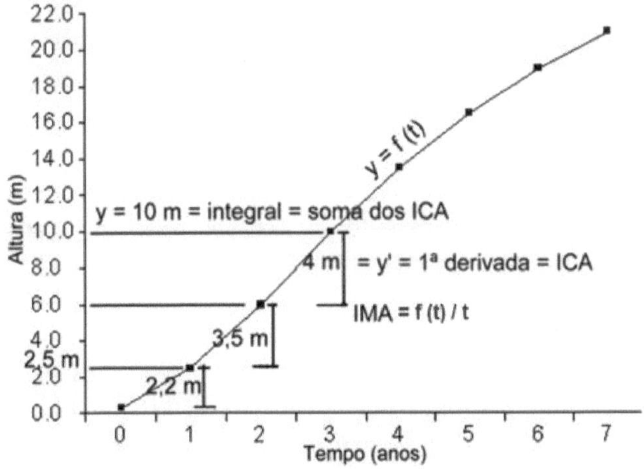

Figura 35 - Graph of a tree's growth in height.

A Figura 36 shows the growth and height increments of a tree.

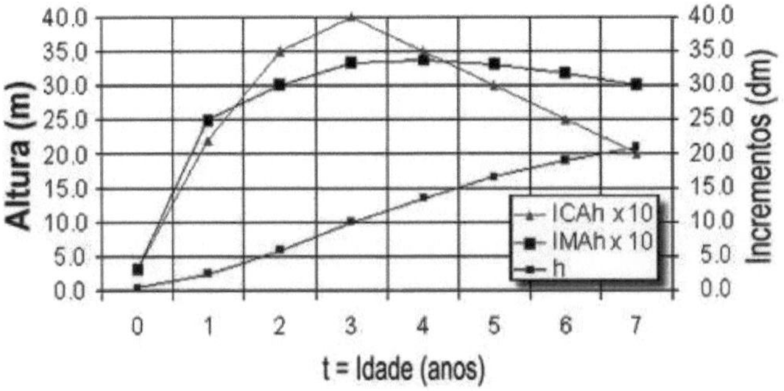

Figura 36 - Growth and height increments of a tree.

4.8 Diameter distribution models

4.8.1 PLANTED FORESTS

NORMAL DISTRIBUTION

The probability density function of the Normal distribution of a variable *x*, with mean μ and variance σ^2, is defined as:

$$f(x) = \frac{1}{\sigma\sqrt{2\pi}} \exp\left(-\frac{(x-\mu)^2}{2\sigma^2}\right)$$

Where: μ = mean, or location parameter; σ = standard deviation, or scale parameter; □□□□x□□□.

The cumulative distribution function is expressed by:

$$F(x) = \frac{1}{2}\left(1 + \operatorname{erf}\frac{x-\mu}{\sigma\sqrt{2}}\right)$$

$$\operatorname{erf}(x) = 2\Phi\left(x\sqrt{2}\right) - 1$$

$$\Phi(x) = \frac{1}{\sqrt{2\,\pi}} \int_{-\infty}^{x} e^{-\frac{1}{2}t^2}\, dt$$

An example of the fit of the normal distribution of diameters is shown in Figura 37 .

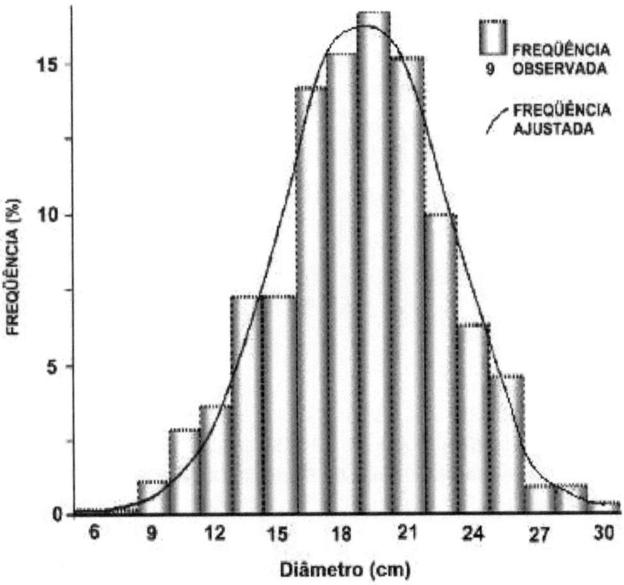

Figura 37 - Normality of the frequencies by diameter class, of the trees contained in the useful area of the plots, at 15 years of age.

LOGNORMAL DISTRIBUTION

A random variable *x* has the Lognormal distribution when its logarithm *y=Ln(x)* has the normal distribution with mean $e^{\mu+\sigma^2/2}$, variance $(e^{\mu^2}-1)\; e^{2\mu+\sigma^2}$ and density function given by:

$$f(x;\mu,\sigma) = \frac{1}{x\sigma\sqrt{2\pi}} \exp\left[-\frac{(\ln(x)-\mu)^2}{2\sigma^2}\right]$$

Where: for $x > 0$, where μ and σ are the mean and standard deviation of the logarithm of the variable (by definition, the logarithm of the variable is normally distributed).

WEIBULL DISTRIBUTION

A variable x with a Weibull distribution has mean *λΓ(1+1/k)*, variance *λ²Γ(1+2/k)-μ²* and a probability density function expressed by:

$$f(x;k,\lambda,\theta) = \frac{k}{\lambda}\left(\frac{x-\theta}{\lambda}\right)^{k-1} e^{-\left(\frac{x-\theta}{\lambda}\right)^k}$$

, *or*

$$f(x;\lambda,k) = \begin{cases} \frac{k}{\lambda}\left(\frac{x}{\lambda}\right)^{k-1} e^{-(x/\lambda)^k} & x \geq 0, \\ 0 & x < 0. \end{cases}$$

Where: for $x \geq \theta$ and $f(x; k, \lambda, \theta) = 0$ for $x < \theta$, where $k > 0$ is the shape parameter, $\lambda > 0$ is the scale parameter and θ is the location parameter of the distribution; when $\theta = 0$, the distribution is reduced to 2 parameters.

The cumulative probability distribution for the two-parameter model is expressed by:

$$F(x;k,\lambda) = 1 - e^{-(x/\lambda)^k}$$

GAMA DISTRIBUTION

A variable x with Gamma distribution has mean *kθ*, variance *kθ²* and probability density function defined by:

$$f(x;\alpha,\beta) = \frac{\beta^\alpha x^{\alpha-1} e^{-\beta x}}{\Gamma(\alpha)} \quad \text{for } x > 0 \text{ and } \alpha, \beta > 0$$

Where: the Gamma distribution is a family of two-parameter continuous probability distributions; it has a scale parameter θ and a shape parameter k; if k is an integer, then the distribution represents the sum of k exponentially distributed random variables, each of which has a parameter θ. Function models for cumulative probability: Linear; Non-linear.

The cumulative probability distribution is expressed by:

$$F(x;\alpha,\beta) = \int_0^x f(u;\alpha,\beta)\,du = \frac{\gamma(\alpha,\beta x)}{\Gamma(\alpha)}$$

BETA DISTRIBUTION

$$f_x(x) = \frac{1}{\beta(a,b)} . x^{\alpha-1} . (1-x)^{\beta-1}$$

Where: $\alpha > 0$e $\beta > 0$then x is defined to have Beta distribution; α, β = parameters of the equation.

The probability density function of the Beta distribution satisfies the differential:

$$f'(x) = f(x) \frac{(\alpha+\beta-2)x-(\alpha-1)}{(x-1)x}$$

The cumulative probability distribution is calculated by:

$$F(x;\alpha,\beta) = \frac{\mathrm{B}(x;\alpha,\beta)}{\mathrm{B}(\alpha,\beta)}$$

$$= \frac{\int_0^x t^{a-1}\,(1-t)^{b-1}\,dt}{B(\alpha,\beta)}$$

$$= I_x(\alpha,\beta)$$

Where: where *B (x ; α , β)* is the incomplete beta function and *Ix* (α , β) is the regularized incomplete beta function.

4.8.2 INEQUITABLE FORESTS

Liocurt realized that the diametric distribution of trees in a natural forest follows a pattern that can be calculated by (SCHNEIDER, 2002):

$$\frac{N_1}{N_2} = \frac{N_2}{N_3} = \dots = \frac{N_{s-1}}{N_s} = q$$

Where: q = Liocurt quotient; N_i = Frequency of diameter class of order i or stratum of order i; i = 1, 2, 3, ... , s = order number of the vertical stratum or diameter class; s = order number of the last stratum or highest order diameter class.

Mayer studied the distribution of diameters in natural populations and found that they have a negative exponential distribution, and the frequency in each class can be determined using the following equation:

$$N_i = K \,.\, e^{-a\,.\,di}$$

Where: N_i = Frequency of the diameter class of order i or of the stratum of order i; "K" and "a" = coefficients of the equation to be estimated for each stand; d_i = value of the diameter class center in cm.

Considering Liocurt's constant q as the ideal frequency ratio between one class and another and Mayer's equation as the ideal diameter distribution model in a natural forest, the Mayer-Liocurt diameter distribution balancing system is obtained as follows (Schneider, 1979):

$$a = ln\, q / h$$

$$q = e^{-b_1 . h}$$

$$ln\, N_i = b_0 + b_1 \,.\, d_i$$

$$K = density / (\Sigma\, (C_1 + C_2 .di + C_3 .d_i^{\,2}) \,.\, e^{-d_i \cdot (ln\, q)/h})$$

$$density = C_1 \,.\, \Sigma\, N_i + C_2 \,.\, \Sigma\, N_i .d_i + C_3 \,.\, \Sigma\, N_i .d_i^{\,2}$$

$$G_i = Y = C_1 + C_2 .d_i + C_3 .d_i^{\,2}$$

$$y_i = b_0 + b_1 \,.\, x_i$$

Where: d_i = DBH class center; q = Liocurt's quotient; h=diameter class interval; N_i = Frequency of class i; Y = G_i = basal area of class i in m^2/ha; b_0 , b_1 , C_1 , C_2 , C_3 = coefficients of the equations; y_i = ln N_i ; b_0 = ln K; b_1 = ln (e^a); x_i = ln (e^{d_i}).

By adjusting the equations of the system, the balanced coefficients of the Mayer equation are obtained, with which the theoretical frequency for each class is calculated. In conclusion, the closer the actual diameter frequency of a species is to the balanced distribution, the more stable the species is in the community.

4.9 Models for trunk shape and tapering

4.9.1 TRUNK TAPERING EQUATION

Due to its versatility in fitting different species, the Prodan polynomial has been used by many authors and has almost always resulted in one of the best models, having been used by Müller (2004) with *Eucalyptus*, Machado et al. (2004) with *Pinus oocarpa* and Scolforo et al. (1998) in a study on *Pinus elliottii*. Prodan's 5th degree equation is expressed as:

$$d_i/d = b + b_{01}.(h_i/h) + b_2.(h_i/h) + b^2{}_3.(h_i/h) + b^3{}_4.(h_i/h) + b^4{}_5.(h/h)_i^5$$

Where: d = diameter at 1.3 m height (cm); h = total height (m); i = position on the stem in relation to the base of the tree; d_i and h_i = diameter and height at position i, respectively.

Some of the models used to estimate the trunk profile are listed in Tabela 20 .

Tabela 20 - Trunk tapering models

NO.	Model	Author
1	$d_i^2/\mathrm{d}^2 = b_0 + b_1\mathrm{h}_i/\mathrm{h} + b_2\mathrm{h}_i^2\mathrm{h}^2$	Munro
2	$ln(d_i/d) = b_0 + b_1.ln[(h - h_i)/h]$	-
3	$d_i^2/d^2 = b_0 + b_1.[h_i/(h-1)] + b_2.[h_i^2/(h^2-1)]$	Kozak *et al*
4	$d_i/d = b_0 + b_1.[h_i/(h - 1{,}30)]$	Munro
5	$ln(d_i/d) = b_0 + b_1.ln[(h - h_i)/h] + b_2.ln^2[(h - h_i)/h]$	Silva and Sterba
6	$d_i/d = b_0 + b_1.(h_i/h) + b_2.(h_i/h)^2 + b_3.(h_i/h)^3 + b_4.(h_i/h)^4 + b_5.(h_i/h)^5$	Prodan

Müller (2004) tested 19 trunk tapering models for *Eucalyptus*, which are presented in Appendix B.

4.10 Production tables

Drawing up production tables depends on the diametric distribution model used, but the process is similar in all cases, with the simplest being the use of the normal distribution.

The process of constructing the production tables with the normal distribution involves the 20 steps listed below, using the equations modelled previously, for each site, individually by age and diameter class.

1) Determination of dominant height (h_{100}) by site and age using the equation;
2) Determination of average diameter (d) by site and age;
3) Determination of the diameter variation coefficient (CV_d) by site and age and the standard deviation (S_d);
4) Determination of diameter class centers (d_i) for 7 classes (i) by age and by site based on multiples of the standard deviation (S) in relation to the average diameter (d), calculated from the smallest to the largest class by:
    - Class 1 - di = d - 1.5 . s;
    - Class 2 - di = d - 1.0 . s;
    - Class 3 - di = d - 0.5 . s;
    - Class 4 - di = d;
    - Class 5 - di = d + 0.5 . s;
    - Class 6 - di = d + 1.0 . s;
    - Class 7 - di = d + 1.5 . s.

5) Determining the theoretical frequency of the class center (ft_i) using the Normal distribution.

6) Determination of the total number of trees per hectare by age and by site, with the actual frequencies per hectare being taken at half the time interval between thinning, due to the fact that the equation estimates by the average. Thus, the number of stock trees at 10 years of age before thinning is the number estimated by the density equation at 8 years and the number of remaining trees is the value calculated by the equation for 12 years. For the stand before thinning at 14 years of age, the value calculated by the equation for 12 years of age is used and for the remaining stand the number of trees calculated for 16 years of age with the density equation is assigned, and so on.

7) Determining the number of trees per hectare (N_i) by diameter class (i) for each site and age with the theoretical frequency (ft_i), having its sum corrected to 100%, calculated by the product of the total number of trees per hectare (N), for the age in question estimated in the previous step, multiplied by the theoretical frequency, using the following equation: $N_i = N \cdot ft_i / (\sum (ft_i \cdot N)$;

8) Determining the average height (h);

9) Determination of the coefficient of variation of average height (CV_h %) by age and by site;

10) Determining the height per diameter class using a similar procedure to calculating the diameter per class according to step 4;

11) Determination of the assortments by diameter class as a function of the diameter and height of the center of each class and the individual volume, which is the sum of the assortments, using the trunk tapering equation and a program developed in the SAS language;

12) Determining the percentage of each assortment per diameter class and the total per hectare, per class and per age on each site;

13) Determination of individual basal area per class and

total basal area per hectare;

14) Determining the number of trees to be cut per hectare in each thinning, by decreasing the stock and forecasting remnants established in steps 5 to 7;

15) Determining the number of trees to be thinned in each class on each thinning occasion by age and site;

16) Determining the basal area per hectare to be thinned in each class and the total per hectare by multiplying the number of trees to be thinned by the individual area in the class, by age and site;

17) Determination of the volume to be thinned by class in a procedure similar to that used for basal area;

18) Determination of thinned assortments by class, by multiplying the proportion of trees thinned by the volume of stock per assortment in each class, by age and site;

19) Determination of the number of remaining trees, basal area and volume per hectare, by reducing the stock existing before thinning by what was eliminated during felling;

20) Determining the remaining assortments by reducing the stock existing before thinning by what was eliminated during cutting.

4.11 Obtaining data for modeling

The type of sampling, the sample size and the frequency of individuals per size class of the dependent variable have a direct influence on the quality of the model to be adjusted or built.

The sample size must be sufficient to allow the model to be adjusted or built and analyzed. It should be borne in mind that the distribution curve of a group of data for a variable (y) only appears when there are a certain number of individuals. Using the Sturges equation ($k = 1 + 3.3 \log_{10} N$) to calculate the number of size classes (k) to use for a given variable in a sample with N individuals, the number of individuals (n_i) per size class (i) can be calculated using the equation $n_i = N / k$. The number of individuals per class n_i must therefore be sufficient to form a distribution curve that can be statistically analyzed. Sturges' rule is considered useful for small samples (less than 200 individuals). Larger samples should make use of rules such as Scott's ($h = 3.5 . s . N^{-(1/3)}$), or Freedman-Diaconis ($h = 2 . IRQ . N^{-(1/3)}$), where h = class range; N = number of observations; s = standard deviation; IRQ = interquartile range. The following empirical rule, developed on the basis of numerous studies with native and planted forests, is applicable to the distribution of diameters:

$k = n^{0,175} . \Delta d^{0,3}$, where k is the number of diameter classes, n is the number of trees and Δd is the range of diameters in centimeters.

Generally speaking, the distribution curve per class is only perceptible with more than 5 observations, becomes better visualized with more than 10 and becomes stable when it approaches 30 observations per class. So, if we have 6 classes, we should consider a minimum sample of 30 individuals, i.e. 5 individuals in each class; we can consider that a number greater than 30 individuals per class would probably be unnecessary, which would give a total of 180 observations. The validation of the best model chosen for the desired purpose, after its adjustment or construction, is what will reveal whether the number of observations used was adequate.

To ensure a good fit and adequate analysis of the residuals, the number of individuals per class must be similar, so that the distribution curves of the residuals per size class of the dependent variable will have the same number of observations.

In sampling, the principle of no human influence in the selection of individuals or units to be sampled and other sampling rules must be respected. Any systematic or random sampling process can be used.

For volume equations or trunk analysis, the point unit system of the nearest tree can be used. To do this, you draw or systematize a number of points equal to the number of individuals you want to show about the population to be studied. The individuals are then divided by size class to find out how many individuals per size class will be drawn and the size of the individual is drawn for each sampling point.

4.11.1 Model fitting and construction

Building or adjusting models involves the following steps:

1) Definition of the dependent variable (research objective);
2) Literature review to find models suitable for the objective;
3) Definition of the independent variables;
4) Definition of the study population and sample;
5) Data collection;

6) Choosing the right software for modeling;
7) Creating the file for processing;
8) Translation of the selected models into the software language;
9) Data processing;
10) Interpretation of results and comparison between models;
11) Choosing the best model;
12) Validation of the selected model.

4.11.2 Model evaluation

Regression analysis is a statistical tool that is used extensively in forestry to adjust and evaluate stochastic models in order to estimate variables that are difficult and time-consuming to measure as a function of variables that are measured more easily and quickly.

Equations can be adjusted using statistical software and spreadsheets based on the simple and generalized least squares methods.

The quality of regression models is directly influenced by the quality of the sample. An adequate sample, with a sufficient number of individuals for the analysis of regression variance and allowing for verification and validation, is essential for assessing their quality.

In the past, researchers indiscriminately transformed the variables in the hope of finding better models, but this practice makes regression analysis unfeasible, as the logarithmic discrepancy distorts the mean square values of the errors and, consequently, Snedecor's F in the regression variance analysis, and it is impossible to estimate the true statistics by approximations, as Meyer and Furnival tried to do. The calculated values are like closed packets and can only be obtained by extracting the logarithm of the dependent variable and redoing the analysis of variance of the regression with the observed and estimated values without the logarithmic transformation.

The terms "model verification" and "model validation" are commonly used to indicate model evaluation. Model validation is one of the phases of model evaluation, involving the process of determining whether the performance of a model is of an acceptable level for its purpose (BURKHART and TOMÉ, 2012).

The simplest evaluation of a model should involve analyzing the variance of the regression, calculating the Coefficient of Determination (R^2) and the Standard Error of Estimate as a percentage of the mean of the observations (Syx%), also known as the Coefficient of Variation, as well as validating it by analyzing the residuals.

Regression models are validated by checking the three assumptions regarding the analysis of residuals: normality of the residuals, homoscedasticity of the variance of the residuals and independence of the residuals. Some authors consider that graphical analysis of residuals may be sufficient for validation, however, there are no universally accepted rules for graphical analysis. Floriano (2004) developed a model for graphical analysis that needs to be better evaluated before it can be used in a generalized way. In the meantime, the use of statistical tests is recommended.

Other complementary tests are used to evaluate models, but are not considered essential, such as: checking for *lack of fit*, checking for *bias*, evaluating *model* efficiency, correlation over time, correlation between components and the use of an independent validation sample.

REGRESSION ANALYSIS OF VARIANCE

A Tabela 21 shows the analysis of variance of the regression of a function $y=f(x_1, x_2, ..., x_n)$.

Fisher's F value, found in the analysis of variance of the regression, gives the value to check whether the regression model under analysis is significant. Statistical software usually reports the probability of significance of F (Pr>F), so there is no need to use F tables. When Pr>F is less than or equal to 0.01, F is "highly significant", or significant at the 1% probability level; when Pr>F is greater than 0.01 and less than or equal to 0.05, F is said to be "significant", being significant at the 5% probability level; and when Pr>F is greater than 0.05, the regression model is "not significant". The significance level can be represented by placing the following symbols in superscript next to the F value:

a) ** - for highly significant F;

b) * - for significant F;

c) ^{ns} - for non-significant F.

Tabela 21 - Regression analysis of variance .

Variation Factor (FV)	Degrees of Freedom (GL)	Sum of Squares (SQ)	Mean Square (QM)	F value
Regression	p-1	$\Sigma(ye-ym)^2$	SQreg/GLreg	QMreg/QMres
Waste	n-p	$\Sigma(y-ye)^2$	SQres/GLres	
Total	n-1	$\Sigma(y-ym)^2$		

Where: y=observed value of the dependent variable; ye=estimated value of the dependent variable; ym=average of the observations of y; GLreg=degrees of freedom of the regression; SQres=degrees of freedom of the residuals; SQreg=sum of squares of the regression; SQres=sum of squares of the residuals; QMreg=mean square of the regression; QMres=mean square of the residuals=variance. Source: Bussab (1986)

STANDARD ERROR OF ESTIMATES $(S\ )_{YX}$

The square root of the Mean Square of the Residuals represents the Standard Error of Estimates (S_{yx}), i.e. the average error of the estimates made with the regression equation in relation to the real observations of the dependent variable. It can be represented as a percentage of the mean of the observations, and is then called the Standard Error of Estimates in Percentage (Syx%), also called the Coefficient of Variation (CV) of the regression.

COEFFICIENT OF DETERMINATION (R^2)

Linear equations

In linear equations, the coefficient of determination is given by the quotient between the sum of squares of the regression and the sum of squares of the total:

$$R^2 = SQ_{reg} / SQ_{total}$$

Where: R^2=*coefficient* of determination; SQ_{reg} =sum of squares of the regression; SQ_{total} =sum of total squares.

Non-linear equations

The non-linear coefficient of determination is given by the difference of unity and the ratio between the sum of squares of the residuals and the sum of squares of the total:

$$R^2 = 1 - (SQ_{res} / SQ\ )_{total}$$

Where: R^2=*coefficient* of determination; SQ_{res} =sum of squares of the residue; SQ_{total} =sum of total squares.

4.11.3 ADJUSTED COEFFICIENT OF DETERMINATION $(R^2)_{AJ}$

When necessary, in the case of a different number of parameters between the best models, or when one model is linear and the other non-linear, or when two linear models are compared and one of them has the constant b_0 and the other does not, the R^2 should be adjusted to enable the models to be compared and the best one to be chosen from the equation:

$$R^2{}_{aj.} = R^2 - \left(\frac{k-1}{N-k}\right) . (1 - R^2)$$

Where: $R^2_{aj.}$ = adjusted coefficient of determination; R = coefficient of determination; *k* = number of coefficients in the equation; *N* = number of observations.

4.11.4 VALIDATION OF REGRESSION EQUATIONS

The tests used to validate the regression equations were carried out using the SAS System statistical package according to the procedures described by SAS (2004).

The selected equations were validated by determining:

- The homoscedasticity of variance using the χ^2 test;
- The independence of the residuals using the Durbin-Watson test or the Breusch-Godfrey test;

- The normality of the distribution of residuals using the Kolmogorov-Smirnov or Shapiro-Wilk tests.

HOMOSCEDASTICITY OF VARIANCE

One of the main assumptions for the usual least squares regression is homogeneity of variance (homoscedasticity). If the model is well-fitted, there should be no pattern to the residuals plotted against the fitted values. If the variance of the residuals is not constant, then the residual variance is said to be "heteroscedastic". There are graphical and non-graphical methods for detecting heteroscedasticity. One graphical method that is generally used is to plot the residuals as a percentage against the fitted values. The SAS System calculates the residuals and fitted values using the GLM, REG and NLIN procedures, which can be displayed on a graph. When the residuals are distributed without any pattern, there is no heteroscedasticity.

A mathematical method for determining whether there is homogeneity of variance in the residuals, which can be carried out using the SAS system, is White's test (SAS, 2004). White's test is computed by finding nR^2 from a regression of e_i^2 on all distinct variables in $X \times X$, where X is the vector of dependent variables including a constant. This statistic is distributed asymptotically as Chi-squared (χ^2) with k-1 degrees of freedom, where k is the number of regressors.

The method tests the null hypothesis that the residual variance is homogeneous. Then, if the "p" value is too small, the hypothesis is rejected and the alternative hypothesis that the variance is not homogeneous is accepted. This is done by using the "SPEC" option in the model statement, as shown in the following example:

```
PROC REG;
  MODEL Y = X / SPEC;
```

The test can also be run using the WHITE option in the FIT statement of the MODEL procedure in SAS, as in the following example:

```
PROC MODEL;
   PARMS A B C;
   Y = A + B * X1 + C * X2;
   FIT Y / WHITE;
```

INDEPENDENCE OF RESIDUES

Durbin-Watson test

The Durbin-Watson test evaluates errors of the AR(1) type. The value of the Durbin-Watson "d" statistic (SAS, 2004) is obtained using the CLM option in the MODEL statement of the SAS GLM procedure, or the DWPROB option in the FIT statement of the MODEL procedure, or the DW option in the MODEL statement of the REG procedure, as shown in the example below:

```
PROC REG;
  MODEL Y=X1 X2 / DW;
```

The "d" statistic is expected to be approximately equal to 2 if the residuals are independent. Otherwise, if the residuals are positively correlated, it will tend to be close to 0 (zero), or close to 4 if the residuals are negatively correlated (Nemec, 1996).

The value of d is given by:

$$d = \frac{\sum_{i=2}^{n}(E_i - E_{i-1})^2}{\sum_{i=1}^{n} {E_i}^2}$$

Where: d = Durbin-Watson "d" statistic; E_i = stochastic error = $\hat{Y}_i - Y_i$; n = number of observations; $\hat{Y}_i$ = estimated value; Y_i = observed value.

Breusch-Godfrey test

The Breusch-Godfrey test considers the possibility of ARMA(p,q) type errors, with errors of order 1 to n. The test is available in SAS as an option in the FIT statement of the MODEL procedure, as in the example:

```
PROC MODEL DATA=DATA;
PARMS A=1 B=1;
Y=A*X**B;
FIT Y / GODFREY=3; *TESTS THE AUTOCORRELATION
OF ERRORS OF ORDER 3;
```

NORMALITY OF THE DISTRIBUTION OF RESIDUALS

The principle of this test is based on comparing the cumulative frequency curve of the data with the theoretical distribution function hypothesized. When the two curves overlap, the test statistic is calculated as the maximum difference between them. The magnitude of the difference is established according to the probability distribution of this statistic, which is tabulated. If the experimental data deviates significantly from what is expected from the hypothesized distribution, then the curves obtained should be equally distant and, by analogous reasoning, if the fit to the hypothesized model is admissible, then the curves develop closely.

The Kolmogorov-Smirnov (D) statistic (SAS, 2004) is an Empirical Distribution Function (EDF) statistic. The Empirical Distribution Function (EDF) is defined for a set of n independent observations X_1 ,... , X_n with a common distribution function F(x). Under the null hypothesis, F(x) is the normal distribution. The observations are ordered from smallest to largest as X_1 ,... , X .n

The empirical distribution function F_n (x), is defined as:

$$F_n (x) = 0, x < X_1$$
$$F_n (x) = i/n, X_{(i)} \leq x < X_{(i+1)} , i = 1,2,...,n-1$$
$$F_n (x) = 1, x_n \leq x$$

F_n (x) is a sequential function that advances by [1/n] with each observation. This function calculates the distribution function F(x). At any value x, F_n (x) is the proportion of observations less than or equal to x, while F(x) is the probability of an observation being less than or equal to x. EDF statistics measure the discrepancy between F_n (x) and F(x). The computational formulas for EDF statistics make use of the transformation of the probability integral U=F(x). If F(x) is the distribution function of x, the random variable U is uniformly distributed between 0 and 1.

Given n observations of x_1 , ..., x_n , the values U_i = F(x_i) are computed as shown below. The Kolmogorov-Smirnov statistic (D) is based on the largest vertical difference between F(x) and F_n (x), and is defined as:

$$D = sup_x |F_n (x) - F(x)|$$

The Kolmogorov-Smirnov statistic is computed as the maximum of D+ and D-, where D+ is the greatest vertical distance between the EDF and the distribution function when the EDF is greater than the distribution function and D- is the greatest vertical distance when the EDF is less than the distribution function.

$$D^+ = max_i \ ((i/n) - U)_i$$

$$D^- = max_i \ (U_i - (i-1)/n)$$

$$D = max \ (D^+ , D^-)$$

SAS System's CAPABILITY procedure uses the modified Kolmogorov D statistic to test the data against the normal distribution with mean and variance equal to the sample mean and variance. In the MODEL procedure, the statistic is only used for samples over 2000 individuals. In the case of small samples, the Shapiro-Wilk test described in section "3.8 Data normality test" is used instead of the Kolmogorov-Smirnov test. In the MODEL procedure, the normality test is obtained using the NORMAL option in the FIT statement, as in the example:

```
PROC MODEL;
  PARMS A B C;
  Y=A+B*X1+C*X2;
  FIT Y / NORMAL;
```

4.12 Analysis of covariance

Analysis of covariance is used to verify the need to use independent functions for the covariates, for example, to describe the hypsometric relationship and volume, where the covariate can be represented by the site index or age. According to Storck and Lopes (1998), when calculating covariance it is possible to check the assumption of homogeneity of the estimated linear regression coefficients

($\hat{\beta}$) by calculating a coefficient for each class separately and then testing their equality.

Schneider (1998) uses the "F" value of the interaction of the covariate with the estimated dependent variable ($\hat{Y}$), calculated by the sum of squares of corrected products, to compare the differences between slopes and uses the "F" value calculated for the covariate through the sum of squares of corrected products of the model without interaction to compare the difference between levels, when there is no difference between slopes. In SAS, the sums of squares of corrected products are obtained using option SS3 of the MODEL statement in the GLM procedure.

5 SITE CLASSIFICATION

The quality of sites can be verified through factors that are intrinsic to the stands or extrinsic.

Intrinsic stand factors are those that relate to the characteristics of the trees and the forest itself, such as:

- Dominant or average height;
- Maximum average growth;
- Total volume at the end of the cycle or rotation;
- Interception;
- Extrinsic factors are those related to the environment, such as:
    a. Biotope factors - climate; lithology; edaphology; morphology;
    b. Biocenosis factors - indicator species (sociology); indicator associations (phytosociology).

Site quality is defined as the capacity of an area for tree growth. It is the response, not development, of a particular species to all environmental conditions. It has been almost universally adopted to represent site quality through quantitative values known as productivity index and site index.

The productivity index is usually expressed as the average increase in annual volume with rotation age. Generally, the site index is expressed as Assmann's dominant height (h_{100}) at a reference age.

The methods for developing site index functions vary according to four factors: the principle applied, the type of data, the construction method and the model used. The most widely used method has been the one that uses height as the site index. Dominant trees are used because they are the least affected by population management and remain in the stand until the end of the rotation. It is necessary to mathematically describe the growth of dominant height by means of functions, in order to establish the classification of sites.

The site index curves of a forest with different site qualities can be similar and proportional, being called monomorphic, or not and present curves of different shapes, being called polymorphic (Figura 38).

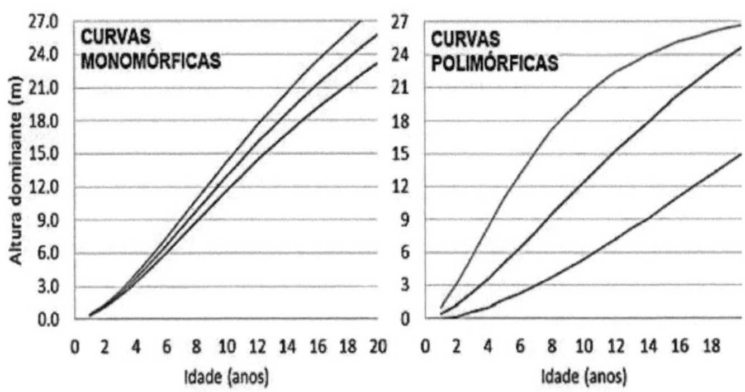

Figura 38 - Shapes of site index curves.

Data to develop site index curves can be derived from:

- Temporary plots - only possible for calculating isomorphic curves;
- Permanent plots - isomorphic or polymorphic curves;
- Trunk analysis - isomorphic or polymorphic curves.

There are three methods for constructing site index curves:

- Guide curve method;
- Difference equation method;
- Nested regression method.

5.1 Guide curve method

This is the most traditional method for constructing isomorphic curves. First, a general equation is adjusted to estimate the dominant height and the dominant height is estimated using the equation for all ages, the curve thus determined being called the "guide curve". Then, an age around 70% of the stand's rotation age is chosen to serve as a reference. At this age, the dominant height is checked and approximated to the nearest even value, which will be the central or base site index (example: if the h_{100} at the chosen age is 20.5m, it is rounded up to 20; if it is 21.2m, it is rounded up to 22). Then assign values to the site indices every 2 meters to cover the entire range of dominant height values occurring at the reference age, starting from the central site index; these will be the Site Indices (SI) in the table. The dominant height values of each site index (SI) at each age (t) are then calculated as a proportion of the dominant heights determined by the equation, as described below.

After determining the IS at the reference age, the values of the dominant heights (h_{dom}) are determined for the other ages at each IS using the equations:

- For ages younger than the reference age:

$$h_{dom(t)} = IS \; . \; h_{100(t)} \; / \; h_{100(t+1)}$$

- For ages older than the reference age:

$$h_{dom(t)} = IS \; . \; h_{100(t)} \; / \; h_{100(t-1)}$$

Where: $h_{dom(t)}$ = dominant height at age t for the IS considered, in meters; IS = Site Index in meters; $h_{100(t)}$ = dominant height of the guide curve at age t; $h_{100(t+1)}$ = dominant height of the guide curve at age t+1; $h_{100(t-1)}$ = dominant height of the guide curve at age t-1; t = age in years.

As an example, consider the following equation adjusted for the dominant height:

$$h_{100} = 40 \; (1\text{-}e \;)^{(-0,07.t)1,7}$$

Calculating the guide curve as in Tabela 22 and choosing 15 years as the reference age, the dominant height of 19.2 meters was obtained at this age, which was approximated to 20 meters, as it was the closest even value to 19.2m, as the central site index. To cover the dominant heights of the forest stands determined in the forest inventory, which have minimum values of 16.5m and maximum values of 23.8m, the site indices were determined every 2 meters starting from the central IS of 20m: 16, 18, 20, 22, and 24, which form the column headings of the IS table and are the same as the h_{dom} values of the IS at the reference age. The $h_{dom(t)}$ values were then calculated for each age t and IS considered.

5.2 Difference equation method

It is used to generate both anamorphic and polymorphic curves. It is assumed that every pair of observations applies to the same height-age relationship. The data is modified algebraically to produce pairs of consecutive differences by:

$$ln(h_{dom_2}) - ln(h_{dom_1}) \quad e \quad I_2^{-1} - I_1^{-1}$$

Where: h_{dom} = dominant height; I = age; indices 1 and 2 = successive observations.

The following equation is fitted by linear regression and is used to generate the other curves:

$$\ln(h_{dom_2}) - \ln(h_{dom_1}) = \hat{\beta}_1 \, (I_2^{-1} - I_1^{-1}) + \varepsilon$$

$$Y = \ln(h_{dom_2}) - \ln(h_{dom_1})$$

$$X = (I_2^{-1} - I_1^{-1})$$

$$Y = \hat{\beta}_1 \, X + \varepsilon$$

Where: h_{dom} = dominant height; I = age; indices 1 and 2 = successive observations. When I_1 is equal to the reference age, h_{dom1} is equal to the site index (S).

Tabela 22 - Example of a site index table .

Age (years)	Guide curve h_{100} (m)	Site Indices - IS (m)				
		16	**18**	**20**	**22**	**24**
1	0,4	0,3	6	6,7	7,3	8
2	1,2	1	1,1	1,3	1,4	1,5
3	2,3	1,9	2,2	2,4	2,6	2,9
4	3,6	3	3,4	3,8	4,1	4,5
5	5	4,2	4,7	5,2	5,7	6,3
6	6,4	5,3	6	6,7	7,3	8
7	7,9	6,6	7,4	8,2	9,1	9,9
8	9,4	7,8	8,8	9,8	10,8	11,8
9	10,9	9,1	10,2	11,4	12,5	13,6
10	12,4	10,3	11,6	12,9	14,2	15,5
11	13,8	11,5	12,9	14,4	15,8	17,3
12	15,3	12,8	14,3	15,9	17,5	19,1
13	16,6	13,8	15,6	17,3	19	20,8
14	17,9	14,9	16,8	18,6	20,5	22,4
15	**19,2**	**16**	**18**	**20**	**22**	**24**
16	20,4	17	19,1	21,3	23,4	25,5
17	21,5	17,9	20,2	22,4	24,6	26,9
18	22,6	18,8	21,2	23,5	25,9	28,3
19	23,7	19,8	22,2	24,7	27,2	29,6
20	24,7	20,6	23,2	25,7	28,3	30,9

5.3 Nested regression method

It can be generated from stem analysis data or permanent plots. It is a hierarchical regression method comprising two types:

- Equations with a common slope;
- Equations with a common independent term.

6 COMPETITION

6.1 Competition indices

Competition indices are an attempt to quantify the influence of neighboring plants on the growth of an individual tree. Many indices have been proposed, but not all can be considered functional; competition indices offer little utility for predicting yield in "typical" stands, and few indices provide better predictions than basal area and other simple measures of stand density and tree classification, which are easier to estimate (VANCLAY, 1992).

Vanclay (1992) reviewed the indices of competition between trees and concluded:

> *"Silvicultural" growth models designed to investigate management options should employ competition indices to provide reliable predictions under extreme silviculture. The need for the model to respond sensitively to extremes of thinning and pruning suggests that indices based on light interception (i.e. both canopy area and sky view) have the greatest potential. Even relatively simple models based on these principles can be useful (VANCLAY, 1992)."*

Therefore, before describing the competition indices, it is worth pointing out that the indices most closely related to growth are those that take into account the basal area of the trunk and the diameter of the canopy. Among the indices that have shown a good correlation with growth are the BAL index (Basal Area Larger), the IOE index (Space Occupation Index) and the living space index (S%), which is calculated as a percentage of the ratio between the average space between trees and the dominant height. Most of the other indices sometimes show some correlation with growth, but few have become operational.

Contreras et al (2011) evaluated 16 measures of tree competition and their effectiveness as estimators of basal area growth in a semi-arid region for three conifer species, finding strong correlations between various competition indices and tree growth. The authors concluded that distance-dependent competition indices were more strongly correlated with growth, with the best of them explaining a greater proportion of growth variation than the best distance-independent index (64% vs. 56%), indicating that competition occurs at the local level. Low correlations between light value indices and growth suggest that in the local semi-arid conditions, trees do not compete primarily for light. The indices used are listed in Annex 3.

Maleki et al (2015) studied the relationship between 18 competition indices and the diameter growth of *Betula pendula* (Appendix 4). The authors concluded that diameter growth is a function of neighborhood interactions and that distance-dependent (spatial) indices were better predictors of growth, but the differences were insignificant in relation to distance-independent (non-spatial) indices; they also found that Ellenberg light values did not contribute to quantifying the effects of competition.

Silveira et al (2015) studied the association between the increase in diameter of *Trichilia claussenii* and competition indices listed in the Tabela 23 .

Tabela 23 - Competition indices used by Silveira et al (2015).

Índice	Fórmula	Tipo	Fonte
BAL (Bal_i)	$Bal_i = \sum_{j=1}^{n_j} G_j$	Independente da distância	Davis (2001)
Glover & Holl (GH_i)	$GH_i = d_i^2 / \bar{d}_j^2$	Independente da distância	Husch *et al.*(2003)
Glover & Holl Modificado (GHM_i)	$GHM_i = d_i^2 / \sum_{j=1}^{n_j} \bar{d}_j^2$	Independente da distância	-
Hegyi (H_i)	$H_i = \sum_{j=1}^{n_j} \frac{(d_j/d_i)}{l_{ij}}$	Dependente da distância	Gadow e Hui (1999)
Spurr (G_{Pi})	$G_{P_i} = \frac{k}{n} \sum_{j=1}^{nj} \left[(i-0{,}5) \cdot \left(\frac{d_i}{l_{ij}} \right)^2 \right]$	Dependente da distância	Husch *et al.* (2003)
Spurr modificado (GM_{Pi})	$G_{P_i} = \frac{k}{n} \sum_{j=1}^{nj} \left[(i-0{,}5) \cdot \left(\frac{\sum_{j=1}^{nj} d_i}{l_{ij}} \right)^2 \right]$	Dependente da distância	-
Razão altura/distância (RHL_i)	$RHL_i = \sum_{j=1}^{nj} h_j / l_{ij}$	Dependente da distância	-

Where: n: number of competing trees; G_j: basal area of competing trees (m²/ha); d_i: DBH of the central tree (cm); d_j: DBH of competing trees (cm); $\bar{d}_j$: average DBH of competing trees (cm); l_{ij}: distance between central tree i and competing tree j (m); DC: crown diameter; k: basal area factor; h_j: height of competing trees; i: order number of the central tree considered; j: order number of the competing tree.

The authors concluded that the diameter growth of the species studied is more closely related to the BAL index. The greater the competition, the lower the growth, showing that the species is strongly affected by competitors. The relationship between growth and the salience index (IS) showed that the larger the canopy, the greater the growth of the species, i.e. *Trichilia claussenii* trees grow better with less competition and a larger canopy. The authors suggest that adjusting the density by means of silvicultural interventions could result in greater growth, with the SI being relevant information for forest management.

6.2 Indices based on the treetops

6.2.1 Cup Competition Factor (CCF)

The determination of the FCC was carried out at a time when access to computers was difficult and calculations with individual trees was almost impossible. It was also tied to the English system of units, which makes comparisons with the decimal system difficult, and was developed based on a simple linear regression to estimate crown diameters, which may not be the best, may be considered an impractical index and may be inaccurate. We therefore suggest using the proportion of space occupied by the projection of the tree crowns (section 6.2.2), which can be determined more precisely and allows comparison with other studies, regardless of the system of units.

The FCC was developed by Krajicek et al. (1961) and reflects the area available to the average tree in a forest stand in relation to the maximum area it could use if it were in open cultivation. To determine FCC values, the ratio between crown diameter and trunk diameter must be established for free-growing trees of the species under study. A function is adjusted to estimate crown diameter (CD) as a function of trunk diameter (d), represented by [CD=f(d)], which is a simple linear regression of the type:

$$DC = b_0 + b_1 . d \ (a)$$

Considering that the crowns of trees grown in open areas have a circular shape, the maximum crown projection surface (MSPC), expressed as a percentage per hectare

(10000m²), occupied by the crown of a tree with a diameter (d), is computed as:

$$MSPC = 100. (\pi/4) . DC^2 / 10000m^2 = 0:007854.DC^2 \ (b)$$

Substituting (a) into (b), we get:

$$MSPC = 0.007854 . (b_0 + b_1 . d)^2$$

$$\square\ MSPC = 0.007854 . (b_0{}^2 + 2b_0 .b_1 .d + b_1{}^2.d^2) \ (c)$$

The FCC of a stand is calculated from a distribution table by diameter class for the stand by adding the MSPC values for each diameter class and dividing by the area in ha. The FCC is expressed by the equation:

$$FCC = (1/A) . (a_0 .\sum n + a_{i1} .\sum n_i .d + a_{i2} .\sum n_i .d_i{}^2) \ (d)$$

Where: FCC = crown competition factor; A = area of 1 ha (10000m²); n_i = frequency of diameter class i; d_i = central diameter of class i; a_0, a_1, a_2 = parameters of the equation to be adjusted.

It is necessary to first adjust a function to estimate crown diameter as a function of trunk diameter for free-growing trees in order to determine the FCC value. Although it is not as widely used as other stand density indices, it has been found to be well correlated with the growth and productivity of various species (BURKHART and TOMÉ, 2012).

Krajicek et al (1961) demonstrated the use of the FCC, starting by adjusting a function to estimate the DC as a function of d for *Picea abies* (Figura 39).

Being:

$DC = 3.12 + 1.826 . d$, where $b_0 = 3.12$ and $b_1 = 1.826$

Substitute b_0 and b_1 into the equation:

$$MSPC = 0.007854 . (b_0{}^2 + 2b_0 . b_1 . d + b_1{}^2 . d^2)$$

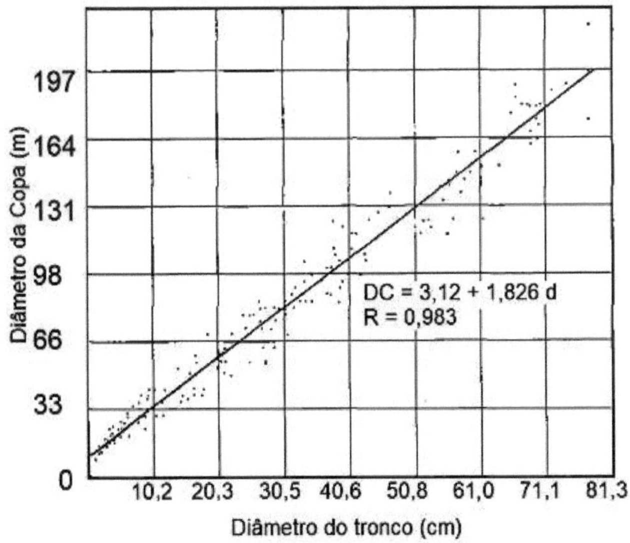

Figura 39 - Function for estimating the crown diameter of *Picea abies* as a function of the trunk diameter of free-growing trees. Source: Krajicek et al (1961).

The equation for estimating the MSPC is$_i$:

$$MSPC_i = 0.007853982 . (9.7344 + 11.41296\, d + 3.345241\, d^2)$$

□ $MSPC_i = (0.076453799 + 0.089637178\, d + 0.026273461\, d^2)$ (e)

With equation (e), the total MSPC is determined for data tabulated by diameter class according to Tabela 24 .

Tabela 24 - Determination of the FCC for an equiaxial stand.

d_i (cm)	n_i (tree/ha)	$n_i.d_i$	$n_i.d_i^2$	Individual MSPC	Total MSPC of class i
1	24,7	24,711	24,711	0,192	4,753
2	17,3	34,595	69,189	0,361	6,241
3	123,6	370,658	1111,973	0,582	71,886
4	155,7	622,705	2490,820	0,855	133,162
5	190,3	951,355	4756,774	1,181	224,801
6	192,7	1156,452	6938,713	1,560	300,701
7	163,1	1141,626	7991,381	1,991	324,762
8	121,1	968,652	7749,218	2,475	299,683
9	76,6	689,423	6204,811	3,011	230,676
10	42,0	420,079	4200,788	3,600	151,236
11	17,3	190,271	2092,981	4,242	73,368
12	7,4	88,958	1067,494	4,935	36,587
13	2,5	32,124	417,608	5,682	14,040
Total	**1134,213**	**6691,608**	**45116,461**	-	**1871,897**

Source: SCHNUR (1937) apud Krajicek et al (1961).

Substituting the sums from the table above into the equation:

$$MSPC = 0.076453799 \sum n_i + 0.089637178 \sum n_i.d_i + 0.026273461 \sum n_i.d_i^2$$

The final result is obtained for the value of the Maximum Crown Projection Surface (MSPC) occupied by the trees in a hectare, called the Crown Competition Factor (FCC):

$$FCC = 0.0764538 \,.\, 1134{,}213 + 0{,}0896372 \,.\, 6691{,}608 + 0{,}0262735 \,.\, 45116{,}461$$

$$FCC = 86.71486945 + 599.8168423 + 1185.365587$$

$$\mathbf{FCC = 1871.80}$$

6.2.2 Space Occupancy Index (SOI)

Based on the same principle as Krajicek et al (1961), using the maximum space occupied by trees growing free of competition to determine the proportion of the ground space occupied by their crown projections and adjusting any regression equation to estimate crown diameter as a function of diameter and other variables that may be necessary, such as height, to obtain an equation with the best statistics, the proportion of space occupied by the surface projection of the tree crowns can be determined.

Considering a function of the type DC=f(d), and that the surface of the crown projection on the terrain is calculated by SPC = π DC² / 4, the proportion of space occupied by the crown projection, or Space Occupation Index by the crown projections of the trees in a sampling unit can be calculated by:

$$IOE = \frac{\sum_{i=1}^{n}(\frac{\pi}{4} \cdot DC_i^2)}{A}$$

Where: IOE = space occupation index by the trees in a sampling unit; DC = crown diameter of tree i, calculated using the DC function$_i$ = f (d$_i$); n = number of trees in the sampling unit; i = number of tree orders in the sampling unit; A = area of the sampling unit.

This index allows comparison with the results of other studies regardless of the system of units used and the equation used to estimate crown diameter, and can be considered a variation of the Crown Competition Factor (CCF) of Krajicek et al (1961).

7 STAND DENSITY

Density is the measure of how much land trees occupy. It can also be referred to as stand or site density and can be represented by various indices, from the simplest such as the number of trees per hectare, to those based on more complex theories.

However, stand variables such as the number of trees per hectare, basal area per hectare or volume per hectare will only be of value for density management if they have a standard for comparison by age and site quality, as they alone do not express the level of forest land cover and use. Density indices based on dendrometric variables are mainly as follows:

- Number of trees (N - trees/ha);
- Basal area (G - m^2/ha);
- Volume (V - m^3/ha);
- Cup Competition Factor (CCF);
- Special indexes (Reineke, Hart-Becking, among others).

The main objective of studying the density of forest stands is to determine which trees to harvest in each period in order to optimize the production of wood for the desired purpose.

Considering density as the occupation of space by trees, the higher the occupation rate, the more green mass there is to promote growth. Up to a certain occupancy rate there will be no competition and the trees will grow freely.

In denser stands, trees grow straighter and more cylindrical. As long as a tree's crown height remains greater than 50% of its total height, there is little influence of density on individual growth; this indicator varies with species, site and age.

When the density becomes excessive due to the growth of the trees and the increased occupation of the ground, it starts to induce the trees to die, starting with the dominated and weaker trees with the smallest canopy. This mortality is called self-thinning.

From a certain occupancy rate, competition is established. The tree trunks begin to thicken less, initially at the base of the smaller trees. As density increases, diameter growth is also reduced in the larger trees and at higher heights. Tree height is less influenced by density, but when it is too high, height growth is also affected.

Extreme competition between plants caused by density can lead to their miniaturization. Slight competition can lead to greater growth in height. With density, fewer and smaller branches are formed.

With thinning, we try to avoid excessive competition, preventing self-thinning and reducing individual growth, but there is a reduction in green mass and therefore growth per unit area.

Determining and controlling the ideal density must take into account the following conditions:

- The trees themselves - species, germplasm and age;
- Environment or site capacity - soil, climate, topography;
- Economic factors - timber market and price, standing timber costs, harvesting costs, transportation costs, etc;
- Availability of labor;
- Wood destination - higher stand densities result in higher wood density and strength, higher production per hectare and smaller trees.

A good density index must have the following characteristics:

- Simplicity and objectivity;
- Absolute and relative quantification;
- Independent of the type of stand, age and quality of the site;
- Correlation with other stand variables;
- Allow quantitative comparisons within and between stands over space and time;
- Make it easier to determine the time and intensity of thinning.

There are many ways of measuring density, each with its own characteristics, but none is considered ideal. The choice of measure to manage density will depend on the species, production objectives, environment and personal preferences. Density can be assessed using simple measurements, density indices or density diagrams:

- Simple measurements: N, G, V, SPC;
- Density indices: Mitchell's D-maximum; Yoda's self-thinning law; Reineke's density index; Wilson's factor; Hart-Becking spacing index;
- Density diagrams: IDS (SDID), DMDP (SDMD).

Density indices seek to combine a biometric variable (d, h, g, v) with the number of trees per ha.

7.1 Population density

It is represented by the number of trees per unit area and is the simplest measure of density.

It is often used to compose indices together with other variables, but it should not be used alone, as it gives no idea of how much space trees occupy - a few large trees can occupy a greater percentage of the space than many small trees.

Population density is calculated by:

$$N = 10000 \cdot n / S$$

Where: N = number of trees per hectare (n/ha); n = total number of trees sampled; S = total area of land sampled.

7.2 Basal area per unit area

It represents the area occupied by the cross-section of tree trunks. It can be used as a density index, as it involves the size of the trees and the space occupied, and is highly related to the surface area of the canopy. When determined in terms of maximum and minimum by age, it allows the thinning rate to be determined. It is calculated by:

$$G = 10000 \sum g / S$$

Where: G = basal area per hectare (m^2/ha); $\sum g$ = sum of the basal areas of the trees sampled; S = total area of land sampled.

7.3 Canopy projection surface per unit area

It represents the area occupied by the cross-section of the tree crowns. It can be used as a density index, as it involves the size of the trees and the space occupied and is highly related to photosynthetic capacity. It allows the thinning rate to be determined by establishing minimum occupancy values per site in order to obtain maximum productivity with the desired wood dimensions.

It is calculated by:

$$SPC = 10000 \sum SPC / S$$

Where: SPC = crown projection surface per hectare (m^2/ha); $\sum$ SPC = sum of the crown projection surfaces of the trees sampled; S = total area of land sampled.

7.4 Volume per unit area

It represents the three-dimensional occupation of the area by tree trunks. It can be used as a density index, as it involves the size of the trees and the space occupied. When monitored, it makes it possible to optimize productivity and establish maximum and minimum limits for the stock at each age, for each site.

It is calculated by:

$$V = 10000 \sum v / S$$

Where: V = volume per hectare (m^3/ha); $\sum$ v = sum of the volumes of the trees sampled; S = total area of land sampled.

7.5 Density indices

7.5.1 WILSON FACTOR

Wilson (1946) considered the average distance between trees to be derived from the relationship between the dominant height and a spacing factor fw:

$$E = h_{dom} \,.\, fw$$

And being:

$$E = \sqrt{(10000/N)}$$

Substituting the first equation into the second gives us

$$h_{dom} \,.\, fw = \sqrt{(10000 / N)},$$

From this we can deduce the Wilson Factor (fw), which is calculated by:

$$fw = 100 / (h_{dom} \,.\, \sqrt{N})$$

Where: h_{dom} = dominant height (m); N = number of trees per hectare.

FW varies with species and stand treatment. It decreases with species tolerance and increases with thinning intensity.

A study in England showed that fw = 0.2 is the appropriate value for roughing grade C/D, which is accepted as the standard, and grades C, D and E would represent 1.5, 0.75 and 0.5 of the density index of the standard grade respectively, i.e. fw values equal to 1.6, 0.23 and 0.28 respectively.

7.5.2 HART-BECKING RELATIVE SPACING INDEX

It is determined by the percentage of the average distance between trees in relation to the dominant height.

It can be used to determine the thinning rate as long as the maximum and minimum values for each age are known.

Older conifer stands on better sites support less relative spacing than younger hardwood stands on worse sites (maximum rates vary between a minimum of 16% and a maximum of 25% depending on the species, the quality of the site and the age of the stand).

The index is calculated by:

$$S\% = 100\ DM / h_{dom}$$

Where: S% = relative spacing index (%); DM = average distance between trees (m) = $\sqrt{(10000/N)}$; h_{dom} = dominant height (m).

The relative spacing index (S%) is shown in the table below as a function of distance and dominant height. The values in light green are those that generally combine higher stand productivity per unit area with larger tree dimensions.

7.5.3 REINEKE'S DENSITY INDEX

Developed in 1933, it is calculated by the number of trees (N) and the average basal area tree diameter (dg) (BURKHART and TOMÉ, 2012).

Reineke obtained a linear relationship between the logarithm of N and dg, using the function:

$$\log N = a \,.\, \log dg + K$$

The density measure based on these two variables is called the Stand Density Index (SDI).

The straight line defined by the equation represents the number of trees/ha in a series of observations for a full-density equianuclear forest.

Reineke (1933) observed that the slope of the line was very similar in various situations, and that a value of -1.605 could be used:

$$log\ N = -1.605.\ log\ dg + K$$

Where: N = number of trees/acre; d = average basal area tree diameter; K = constant that varies with the species.

Reineke showed that by taking the logarithm of the number of trees (N) on the ordinate and the root mean square diameter (dg) on the abscissa, a linear relationship is obtained between the variables.

The index is independent of age and site and expresses the number of trees per hectare for an equiaxial stand at full density.

7.6 Density control

Density control throughout the life of a forest stand is carried out by means of intermediate cuts between the planting of the forest and the end of the rotation, known as thinning.

7.6.1 Planting density

Planting density depends on the size of the wood to be obtained and the size of the machines used for maintenance and harvesting. Thicker logs are obtained with wider spacings, but productivity per hectare is higher with smaller initial spacings, as the canopies close earlier and carry out more photosynthesis. Maintenance machines are usually less than 2.5 m wide and can cross gaps between 3 and 3.5 m with some ease, while harvesting machines have limited reach for selective thinning. Planting spacings of 3 meters between rows and 2 to 3 meters between plants in the row are common for fine wood production. For wood destined for sawmills and lathes, there are companies that adopt spacings of 4 to 6 meters between rows and 2 to 3 meters between plants in the rows.

Experimenting with planting spacing is a desirable practice in order to define the best one for obtaining the desired products.

7.6.2 Thinning

These are intermediate cultural cuts carried out in immature stands to encourage the growth of trees that will remain in the stand for longer, with the aim of obtaining larger timber with greater added value.

Objectives of thinning

The main objectives of thinning are:

1) Use or sell trees that could die and decompose;
2) Redistribute the growth of wood from lower quality trees to higher quality trees, increasing the value of the wood from the remaining trees;
3) Increase the profitability of growing trees by reducing investment in full volume;

4) Provide money to pay for investments, with the income from thinning;
5) Improve non-timber uses of the forest, such as grazing, wildlife and recreation;
6) Obtain more frequent periodic rentals.

TYPES OF THINNING

They can be of three main types:

1) Selective - trees of certain dominance classes are cut down;
2) Systematic - entire rows of trees are cut down;
3) Mixed - a combination of systematic and selective.

Selective thinning from below - the smallest trees in the stand are cut down. Example: thinning from below in an original 3m x 2m plantation, or 1667 trees/ha, with the top 950 trees/ha (57%) remaining. Some points are more concentrated than others when thinning from below, so it is necessary to control the spaces between the trees when marking the thinning. The quality of the marking depends on the skill and experience of the markers. The result of thinning from below is stands with more space for the remaining trees, which are larger and therefore more valuable.

Systematic thinning - entire rows of trees are eliminated on each occasion; they can be programmed in different ways, for example: cutting the 5th row in the first thinning to form skid trails; cutting every second row on a single occasion; cutting the 13th row in the first thinning, the 5th and 9th rows in the second and the 3rd, 7th and 11th in the third; cutting the 6th row in the first thinning and the 3rd row in the second; etc.

Usually, a mixed cut is carried out on the first occasion, eliminating the 5th planting row, since the machines (harvester and feller-buncher) can reach two rows away on each side, allowing a complementary selective thinning to be carried out on the rest of the intermediate rows - the systematic cut, in this case, also aims to facilitate the transport of the wood. On the second occasion, only one more selective cut can be made from below;

7.7 Effects of thinning on the forest and trees

Photosynthesis is the process that stores energy for trees and is controlled by the availability of:

- Temperature;
- Sunlight;
- Water;
- Nutrients;
- Green mass of the tree (amount of chlorophyll).

Temperature controls respiration, which uses food to maintain the tree's life processes and release the energy needed for growth. The rate of growth depends on how much energy and food is left over after the tree has used what it needs for respiration.

Removing some trees and reducing competition leaves more nutrients and water for the remaining trees. The opening up of the canopy increases the availability of light for the remaining trees. This increases carbohydrate production. Initially, after thinning, there is an acceleration in diameter growth. Normally, there is a greater acceleration in diameter growth than in height. Height growth will only be affected if the previous density has reached levels that are detrimental to height growth.

The effect on tree growth varies from species to species and with the age of the trees. Older trees (after the culmination of ICA) respond less to thinning, particularly heliophilous species. Shade-tolerant species can grow in the shade and respond more to thinning at older ages. In some cases, thinning causes growth to slow down and even trees to die. This undesirable reaction to thinning is called "thinning shock". It occurs in trees with small crowns that are less than 1/3 of the tree's total height.

Repeated thinnings stimulate vigorous growth and promote the formation of canopies with more than 40% of the tree's total height. The shock caused by thinning is unlikely in such stands.

Bertoloti et al (1983) studied the effects of thinning on tree growth in a thinning experiment with *Pinus caribaea* with the following treatments:

1) Mechanical thinning (shallow cutting) on the entire 3rd row;
2) Thinning from below (removal of lower trees);
3) Mechanical roughing on the entire 5th row + roughing from below (Combined method);
4) Mechanical roughing on the entire 9th row + roughing from below (Combined method);

5) Mechanical roughing on the entire 13th row + roughing from below (Combined method);
6) Mechanical grinding on the entire 17th row + grinding from below (Combined method);
7) Mechanical grinding on the entire 19th row + grinding from below (Combined method);
8) Witness (no thinning);
9) Thinning down -30% of the lower trees every 2 years.

The authors concluded that, considering the results obtained up to 14 years of age: the first thinning can be carried out at 11-12 years of age for the species *Pinus caribaea* var. *hondurensis* planted at a spacing of 2m x 2m, in Agudos - SP; thinning had the effect of considerably increasing the number of individuals with a diameter greater than 25 cm in relation to the control up to three times, depending on the treatment, but without increasing total production in terms of wood volume.

Thinning always reduces total wood production, as the area occupied by the canopy is reduced by the operation, reducing growth per unit of forest area. Coarse wood production is increased, as the remaining trees grow more vigorously after thinning.

7.8 Thinning criteria

The first thinning should be carried out before the maximum ICA in diameter, while the trees are well formed; after that, they will have lost vitality and will not be able to take full advantage of the increased space.

The growth of the remnants is greater when thinning is earlier and the canopies have not yet completely closed.

When the first thinning is early, the rotation period will be shorter.

Total production will be higher the less intense the thinning.

The stronger the thinning and the longer its cycle, the greater the proportion of stock removed.

7.8.1 Criteria for the 1st thinning

- A sufficient amount of wood must be thinned to be economical - the later and the greater the intensity, the better;
- It must not affect the stability of the stand - the later and more intense, the worse (winds);
- It shouldn't be too late that it reduces future individual growth;
- It should be carried out at the age of maximum ICA in DAP;
- It should be carried out when the S% drops to 16% for *Pinus elliottii* and before it reaches 18% for *Eucalyptus* spp;
- The intensity must be such that the stand closes its canopy again within the period foreseen until the next cut.

7.8.2 Choosing the trees to thin

- Health - healthy trees are preferred to remain;
- Stem quality - good quality trees should remain;
- Dimensions - small trees are preferred for thinning;
- Spatial distribution - avoid excessive local density.

7.8.3 Thinning Considerations

On the one hand, enough wood must be cut to be economical, and on the other hand, the percentage cut must not affect the stability of the stand or the future increment (productivity per hectare);

It should be combined with adequate planting spacing for the forest's objectives and to provide as few consecutive thinnings as possible, as these are generally low-profit operations

Sometimes the first thinning is considered to be pre-commercial, as its main purpose in this case is not to produce wood.

7.8.4 Thinning markings

The size of the trees to be thinned depends on the expected growth of the remaining trees; it can be determined by various methods, such as the one that uses spacing relative to the dominant height.

The choice of each tree to be thinned is determined taking into account the following priority criteria:

- Tree size (cutting diameter limit and dominance class);
- Vitality and health;
- Quality and shape of the trunk;
- Spatial distribution of trees.

7.9 Thinning calculation

Some of the most widespread methods are the following:

- Mexican - uses the compound interest formula applied to growth;
- English - cut 30% of the expected growth volume until the next thinning;
- Basal Area - uses desired maximum and minimum basal area limits per hectare per age (based on experiments or local data);
- By S% - uses the Hart-Becking relative spacing index;
- Croatian - similar to Mexican;
- Abetz - uses the h/d ratio as a reference for thinning;
- Wilson - uses the relationship between spacing, dominant height and the Wilson spacing factor.

7.9.1 Thinning calculation using the Hart-Becking Index

Dominant height (h_{100}): the average height of the 100 thickest trees per hectare.

Relative spacing (S%): is the ratio, expressed as a percentage, between the average linear distance (MS) between trees and the dominant height ( ).$_{h100}$

Thinning from below determined by relative spacing (Hart-Becking Index):

- Relative spacing of 24% is too open;
- Relative spacing of 16% is too close;
- Thinning should be carried out when the S% is between 16% and 18%, increasing to 21% to 24%;

- Older stands can withstand greater density.

7.9.2 English thinning method

The English method (SCHNEIDER, 1993) determines an annual cutting rate of 70% of the maximum IMA.

In order to know the maximum Average Annual Growth Rate ($IMA_{máx}$), it is necessary to use a production table, because at the age of the first thinning, it has not yet been reached.

A maximum thinning weight of 40% of the volume is also recommended, due to the risk of wind damage and excessive reduction in the annual increment after thinning.

The cutting rate (TC) is calculated by:

$$TC = 70 * IMA_{máx} / 100$$

Example of calculation using the English method

Consider a forest with the following characteristics:

- Current stock = 320 m^3
- $IMA_{máx}$ = 35 m^3/ha/year
- TC = 0.7 . 35 = 24.5 $m^3 . ha^{-1} . year^{-1}$
- Thinning interval = 5 years

The cutting rate (TC) and thinning weight (P) are calculated as follows:

$$TC = 24.5\ m^3.ha^{-1}.year^{-1} .\ 5\ years = 122.5\ m^3/ha$$
$$P = 100\ .\ 122,5 / 320 = 38,28\ \%$$

7.9.3 ABETZ THINNING METHOD

Abetz developed several selective thinning methods, in which the trees to remain until the end of the rotation are chosen at the time of the first thinning. The Abetz method described by Schneider (1992) uses the h/d ratio as a reference for thinning, considering a h/d = 1 ratio as balanced; when the value of the h (m) / d (cm) ratio stops:

- = 1 - growth in height and diameter is considered balanced;
- > 1 - the height is growing proportionally more than the diameter and there is a need for thinning;
- < 1 - the diameter is growing more than the height and there is little competition and generally no need for thinning;
- < 0.9 - in this case, if thinning is carried out, its weight should not be greater than 40%.

There is no need to determine the age and site index. Cannot be used when the diameter distribution is irregular and wide-ranging. It is not suitable for specific objectives, such as achieving a certain desirable diameter in a given period.

The h/d ratio indicates the level of competition and also the vitality, growth potential and stability of the trees against the wind.

7.9.4 WILSON SPACING FACTOR

This criterion was originally introduced by Wilson (1946), by which name he renamed it and its index.

According to this index, the number of trees per unit area (N) can be obtained as a function of the dominant height

(h_{100}) of the stand and a factor f, average spacing, expressed as a percentage of that same stand height.

Considering a square distribution, with the average distance between trees $e = h_{100} . f$, N, the number of trees per hectare, becomes:

$$N = 10,000 / e^2$$

The Wilson Factor (WF) is deduced from this expression:

$$FW = 100 / (h_{100} . \sqrt{N})$$

This stand density index is independent of the age and quality class of the site. According to the author, fw varies with species and stand treatment. This index decreases with species tolerance and increases with thinning intensity.

In a study carried out in England, it was considered, based on experiments carried out on thinning, that fw = 0.2 would be the appropriate value for the C/D thinning grade, accepted as the standard, and that grades C, D and E would represent, respectively, 1.5, 0.75 and 0.5 of the density index of the standard grade, i.e. fw values equal to 1.6, 0.23 and 0.28, respectively.

The table on the next page shows the numerical values for dominant heights ranging from 8 to 36 meters, degree of thinning C, C/D, D, E. and Wilson index of 0.16; 0.20; 0.23; 0.28, respectively.

Table of the number of trees/ha according to dominant height and degree of thinning. Source: Monteiro Alves (1982) apud Schneider (2009).

7.9.5 MEXICAN THINNING METHOD

It refers to the annual volumetric increments of a tree, which accumulate following the law of compound interest extended to forest stands. The thinning volume is a function of the existing stock before and after the cuts and the ICA_v before the cuts; it allows the cycle and weight of the thinning to be determined.

It makes it possible to determine cuts according to growth, with variable intensities, for each stand subject to sorting, with a fixed interval (cutting cycle), during which the forest stand responds to the increase in the remaining trees, establishing rotations that tend to be repeated indefinitely.

MATHEMATICAL EXPRESSIONS:

a) Possibility of an annual cut (PC):

$$PC = VC / cc$$

b) Actual stocks before cuts (ER):

$$ER = VP.\ 1{,}0p^{cc}$$

c) Cutting cycle in years (cc):

$$cc = (\log ER - \log VP) / \log 1.0p.$$

d) Cutting intensity (CI):

$$IC = [1 - (1 / 1.0p^{cc})].\ 100$$

7.9.6 CROATIAN" THINNING METHOD

It was developed by Klepac (1980/4) apud Schneider (1992); it is similar to the Mexican method, but the cutting intensity is multiplied by a thinning factor.

Under this method, in normal forest, final and early production are not distinguished separately, so the cutting rate represents the difference between the normal volume of the stand before and after cutting.

MATHEMATICAL EXPRESSIONS OF THE CROATIAN METHOD

$$TC = V - VN$$

$$VN = V / 1.0p^{cc}$$

$$TC = V. \{1 - (1 / 1.0p^{cc})\}$$

Where: TC = cutting rate; V = normal volume, before selective cutting; VN = normal volume, after selective cutting; p = percentage of increment; cc = cutting cycle; TC = cutting rate.

The removal by thinning is only part of the increment and not the total accumulated increment removed by the Mexican method, calculated by:

$$TC = \{VN . (1.0p^{cc} - 1). 1/q\} / (VN. 1.0p^{cc}) .100$$

Where: 1/q = thinning factor, varies according to circumstances: in young stands it is 1/4; in medium-aged stands it is 1/3; and in old stands it is 1/2 or greater; p = percentage of increment.

7.10 Final thoughts on density and thinning

The higher the density of the trees in the stand, the following results:

- Lower growth in individual tree diameter and volume;
- Greater growth in height up to a maximum point, when it decreases with increasing density;
- Increased growth in volume per hectare up to a maximum point, when it is reduced;

- The trunks are more cylindrical, thinner and have a higher wood density.

Controlling tree density is one of the most important tasks in forest management with thinning, as it has a major influence on individual productivity and total production.

Thinnings have a direct influence on revenue and production costs; light thinning in large numbers has reduced revenue and increases total thinning costs.

Density and thinning influence the stability of stands in relation to winds and resistance to pests and diseases. Thinning too heavily reduces wind resistance and can facilitate the occurrence of pests and diseases.

8 PRUNING PLANTED FORESTS

Due to the relationship between the pruning of cultivated trees and the density of stands, it was decided to associate it with this chapter.

The process of natural pruning is slow and lateral branches persist in some species, forming knots that are usually dry and loose, depreciating the wood. More widely spaced stands provide larger timber. However, trees that grow more freely have thicker and more numerous branches. One of the techniques for reducing the size and number of lateral branches is to increase the planting density. However, it has been observed that trees growing in dense forest stands have a tendency towards imperfect natural pruning, with the presence of a single branch being enough to depreciate several pieces of wood obtained from a log; in addition, individual growth is reduced due to the high competition between trees, and in the end, smaller logs are obtained, reducing the yield in sawmills and rolling mills.

The method used to prevent the formation of knots in the wood of Pinus species is pruning or artificial pruning, which is considered an expensive practice but is generally compensated for by the higher value of the wood produced. However, it has been found that production is influenced by the intensity with which pruning is applied; therefore, determining the maximum intensity with which it should be carried out is of paramount importance for maximizing the production of knot-free wood. The quality of the wood is highly influenced by treatments such as pruning, an operation that reduces the productivity of the trees, but which is necessary in the production of wood for noble use such as carpentry, considerably increasing its value.

Artificial pruning is the cutting of the lower branches of the crowns of trees grown for timber production, with the aim of obtaining knot-free wood, of better technological quality, more homogeneous and with a better appearance.

When choosing the diameter for the corrugated core (Figura 40) the product to be obtained must be considered, as follows:

- Wood for sawmills - as little as possible;
- Lathe lamination - equal to the limit that the machine (lathe) can reach (usually 8 to 11 cm);
- Parallel lamination (beveled) - as little as possible;
- Fibers and particles - as little as possible.

As the trees grow, competition increases and the availability of light, nutrients and water decreases for each individual tree, the lower branches die and, despite the decrease in individual tree growth, the volume stock continues to increase and the radial growth of the trunk is reduced in the lower parts (KRAMER AND KOZLOWSKI, 1972), resulting in a more cylindrical shape. This makes pruning important, as it eliminates live and dead branches from the trees, providing a greater quantity of wood without knots and, consequently, of better quality and greater commercial value (GIBSON *et al.*, 2001), as well as providing easier access and visibility through the forest stand (WARNER, 1997).

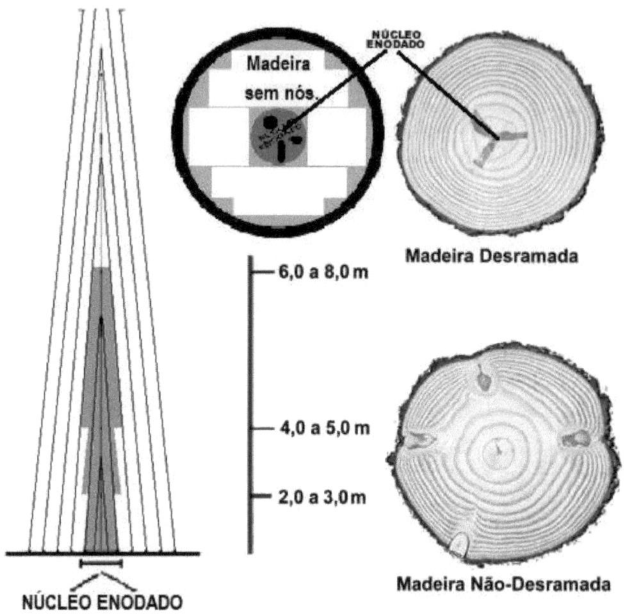

Figura 40 - Enodized core, pruned wood and unpruned wood.

Floriano (2004) studied pruning treatments with the aim of determining the influence of the intensity of pruning on the growth and shape of *Pinus elliottii* in a plantation from 6 to 15 years old in Piratini, southeastern Rio Grande do Sul, using four treatments represented by the percentages of pruning: 0% (control without pruning) and 40%, 60% and 80% pruning in relation to the total height of the trees. The treatments began when the trees were 6 years old and were repeated every year until they reached a height of 6 meters in all treatments. The results, when the plantation was 15 years old, showed that there was no significant influence of pruning on average height, dominant height and mortality; diameter and volume without bark were significantly influenced by pruning, with the following reductions compared to the control treatment without pruning: 2.0% of diameter and 3.5% of volume in the 40% pruning treatment; 7.5% of diameter and 15.0% of volume in the 60% pruning treatment; 13.5% of diameter and 26.9% of volume in the 80% pruning intensity treatment. Pruning significantly influenced the shape of the trees, with the pruned treatments having a conical shape with a slight tendency towards paraboloid and the control treatment with no pruning having a conical shape with a slight tendency towards neiloid. The percentage of clean, knot-free wood obtained in the three pruning treatments was similar, reaching an overall average of 46% of the total individual bark-free volume produced up to 15 years of age.

Wood can have three types of knots: live knots, dry knots attached to the wood and loose dry knots. The latter are the most damaging, causing holes in sawn or laminated timber.

The main benefits of pruning are: clean, knot-free wood; increased revenue and profit from wood; easier penetration and visualization; reduced thinning costs; reduced use of chemicals in pulp production; improved quality of fibre and particle mass in board production; reduced waste in wood processing and increased productivity.

The consequences of pruning are: reduced growth (Figura 41); changes to the shape of the trees; an increase in the forest cover with greater fire intensity; and an increase in the population of certain pests and diseases that use the forest cover to feed and spread.

The first pruning is usually carried out on 100% of the trees in the stand with the aim of facilitating mechanized operations inside the stand and improving the quality of the wood. The operation is expensive and in subsequent prunings it is not justified to prune all the trees, as the smaller ones will certainly be eliminated in the first thinnings and will not have time to accumulate a significant layer of knot-free wood to justify the operation.

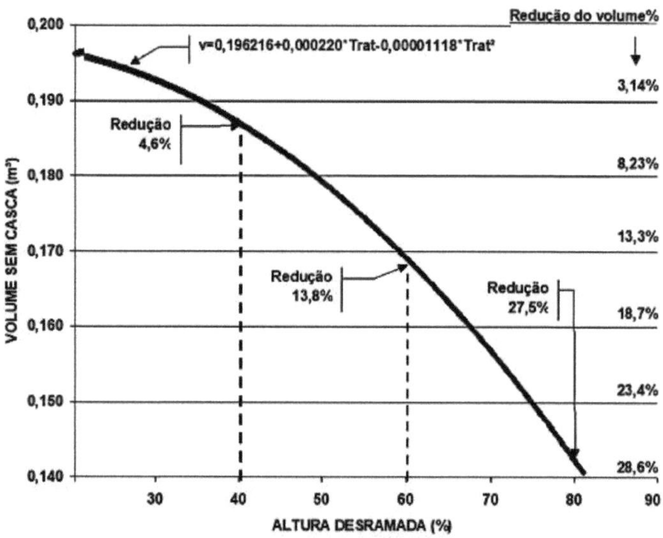

Figura 41 - Reduced growth of *Pinus elliottii*, at 15 years of age, caused by pruning started at 6 years of age, in Piratini, RS.

Pruning is usually done by hand and is carried out to a height of 2.5 m on the first occasion, 5 m on the second occasion and 7.5 m on the third. The planning of pruning must take into account the maximum diameter desired for the core of the trunk. Considering that a laminating lathe has a limit of laminating up to 10 cm in diameter, when the base of the logs to be obtained from the trees intended for pruning reaches this dimension, pruning is carried out. This can be planned using an equation for diameter growth of the type [d=f(t)], trunk tapering equations [di=f(d,hi/h) and hi=f(h,di/d)] and an equation for diameter distribution by age and site [ni=f(d,IS,t)]. With the diameter distribution at the desired age, the trees to be thinned are chosen, taking into account those that will be able to accumulate enough knot-free wood before being felled, and the height at which they will present the maximum diameter for the cored core at each age and diameter class is determined. In short: on the first occasion, at around 5 years of age for *Pinus* and around 1.5 years for *Eucalyptus*, all the trees are pruned; on the other occasions, only the trees that will remain in the stand for another 5 to 8 years.

When planning pruning, the following general rules should be considered, but they may vary depending on the case:

- Pruning should not exceed 40% of the tree's total height from the ground, to avoid a significant reduction in growth;
- The maximum diameter of the core varies depending on the future use of the knot-free wood, generally between 8 and 11cm;
- 1st pruning - should be carried out up to about 2.5 m in height, starting when the trees are about 6 to 6.5 m high; the diameter, at about 20 cm from the ground, of trees with above-average DBH, should not exceed the limit diameter of the pollarded core;
- 2nd pruning - should be carried out up to a height of around 4.5 m. It should be started when the

diameter of the trees to be pruned at a height of 2.5 m has reached the diameter limit for the core;

- 3rd pruning - this should be carried out up to a height of around 6.5 m. It should be started when the diameter of the trees to be pruned at a height of 4.5 m has reached the diameter limit for the core;
- Pruning should be carried out in the cooler season, when the plants are at rest and suffer less stress;
- Pruning should only be carried out in the warmer season when it is extremely necessary in plots with delayed pruning, as long as it does not involve removing green branches;
- The proper ages for pruning vary according to the site and species; normally in Pinus, the 1st, 2nd and 3rd pruning ages are carried out at 4-6, 6-8 and 9-11 years, respectively, varying according to the growth of the trees.

9 ECONOMIC AND FINANCIAL ANALYSIS

According to Rezende and Oliveira (2001), the economic evaluation of a project is based on its cash flow, represented by the costs and revenues distributed over the useful life of the project, resulting in a net income for each period considered, obtained by the difference between revenues and costs associated with the project. The author suggests the following criteria for the economic and financial analysis of forestry projects:

- Present Value (PV) or Net Present Value (NPV);
- Internal Rate of Return (IRR);
- Equivalent Periodic Cost (or Benefit) (CPE or BPE), also known as Equivalent Annual Value (EAV);
- Average Cost of Production (CMPr);
- Benefit/Cost Ratio (B/C);
- Expected Land Value (ELV), or Soil Production Value (SPV);
- Capital Recovery Time.

An investment project can be represented schematically by a flow of monetary income and expenditure, estimated at the end of each period of time (FARO, 1979), as in Figura 42 which illustrates the cash flow of a conventional project, where positive net income is represented by upward arrows and negative net income by downward arrows.

There are several factors that can influence the economic cutting age, such as: the planning horizon, the costs involved in production, the price of wood, the productivity of the site and the discount rate used (SOARES et al., 2003). Investing in a project means investing resources over a period of time, with the expectation of obtaining products in subsequent periods.

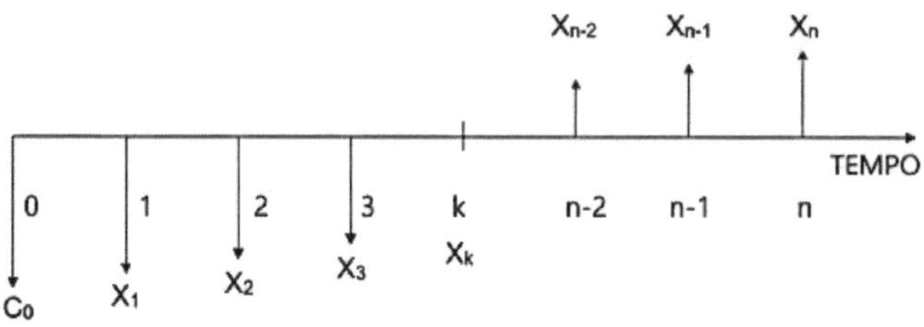

Figura 42 - Generic cash flow diagram for a conventional investment project. Source: Rezende (2005).

It only makes sense to analyze one or a sequence of investments within a given period of time or Planning Horizon (PH). Faro (1979) defines the PH as the estimated period of time during which the enterprise under analysis will operate. For Solomon and Edin (1977), the Planning Horizon is the period of time in which one is prepared to make forecasts and calculations for future plans. Johnston *et. al.* (1977) consider the PH to be the relevant period of time for analyzing an activity, i.e. not so long that it becomes mere speculation and not so short that it does not allow the evaluation of its economic, social and environmental results.

Forestry economic evaluations are generally carried out within the concept of forestry investment projects, which is understood as the investment of capital in a given enterprise, with the aim of obtaining income, within a planning horizon equal to a rotation for the most common use of the wood to be produced.

The economic evaluation starts by determining income and expenses, considering a discount rate obtained from the interest rate practiced on the financial market for the type of enterprise considered.

Revenues can be calculated by:

$$R_r + \sum_{j=1}^{r} Dj * 1{,}0i^{r-j} + N_q * 1{,}0i^{r-q}$$

Where: R_r = value of the final cut at rotation age (r) per hectare; D_j = thinning revenues at age j per hectare; N_q = revenues from non-timber products, after discounting production costs, at age "q" per hectare; r = rotation, in years; j = thinning age.

Expenses are determined by:

$$C * 1{,}0i^r + A * (1{,}0i^r - 1)$$

Where: C = production costs; i = discount rate; A = administration cost.

9.1 Net Present Value (NPV)

The NPV is defined as the equivalent amount, at date zero, of a financial flow, discounting the interest rate determined by the market. It is obtained by updating income and expenses up to the year the project is implemented, discounting the updated expenses from the updated income each year (JOHNSTON et al., 1977), i.e. calculating the present value of a future cash flow, which can be expressed by the formula (SCHNEIDER, 2002):

$$VPL = \left[\frac{Rr + \sum Dj.1{,}0i^{r-j} - C.1{,}0i^{r} - \frac{a+t}{0{,}0i}.(1{,}0i^{r} - 1)}{1{,}0i^{r}} \right]$$

Where NPV = net present value; Rr = revenue from the final cut at age r; Dj = revenue from thinning carried out in year j; C = implementation cost; a = annual administration cost; r = rotation, in year; t = annual lease cost; i = annual interest rate.

According to Johnston et al (1977), the NPV is more accurate when it is calculated for a single rotation of the crop in question, taking into account reductions in income due to natural phenomena that affect the stands. If the present value is positive, the investment proposal is attractive, and the higher the positive value, the more attractive the proposal (PAMPLONA and MONTEVECHI, 1999).

9.2 Benefit/Cost Ratio (B/C)

The ratio of benefits to costs (B/C) becomes an indicator of economic and financial efficiency as it suggests the return on investment based on the relationship between total revenue and the expenses incurred to make it possible (DOSSA et al, 2000).

The benefit/cost ratio is determined by dividing the present value of revenues by the present value of costs (WILLIAMS, 1988). The value of the Benefit/Cost ratio is obtained using the formula (SCHNEIDER, 2002):

$$B/C = \left[Rr + \sum Dj.1{,}0i^{r-j}\right] / [C.1{,}0i^{r} + (a+t)/0{,}0i.(1{,}0i^{r} - 1)]$$

Where: B/C = benefit/cost ratio; Rr = revenue from final cut at age r; Dj = revenue from thinning carried out in year j; C = implementation cost; a = annual administration cost; r = rotation, in year; t = annual lease cost; i = annual interest rate.

This is normally a method used by government agencies when comparing public projects. This is because the decision on the best project alternative is made for the project with the highest B/C ratio (SCHNEIDER, 2002).

9.3 Internal rate of return (IRR)

The minimum rate of attractiveness of capital is understood to be the average remuneration being paid in the economy for each monetary unit invested in it, plus an additional gain that must accompany the business capacity and risk (DOSSA et al, 2000).

The internal rate of return (IRR) makes it possible to compare the profitability of different projects, forestry and agriculture, for example. The IRR is represented by the profitability generated by the project itself, and is independent of arbitrarily chosen interest rates. It is defined as the interest rate that equals the present value of future expected returns with the present value of future expected costs (WILLIAMS, 1988), i.e. it is the interest rate at which the net present value (NPV) is zero (NEVES et al, 2001).

Thus, the IRR makes it possible to check whether the cost of capital is greater than, equal to or less than the profitability of a given project, making it possible to assess its economic viability.

Algebraically, the internal rate of return is the real, non-negative discount rate *i* for which the relationship is verified (SCHNEIDER, 2002):

$$\left[Rr + \sum Dj.1{,}0i^{r-j}\right] = [C.1{,}0i^{r} + (a+t)/0{,}0i.(1{,}0i^{r} - 1)] = 0$$

Where *Rr* = revenue from the final cut at age *r*; *Dj* = revenue from thinning carried out in year *j*; *C* = implementation cost; *a* = annual administration cost; *r* = rotation, in years; *j* = realization age in years; *t* = annual lease cost; *i* = annual interest rate.

The decision rule indicates that it will only be advantageous to invest in the project under consideration if the calculated IRR is higher than the interest rate on the financial market. The choice of an investment must necessarily fall on the one with the highest IRR. The higher the IRR, the more desirable the investment (DOSSA et al, 2000).

9.4 Annual Equivalent Value (EAV)

The equivalent annual value is the annual expression of the net present value over the computed planning horizon, at a given discount rate. The EAV is an extremely useful criterion for comparing investments with unequal periods or horizons (MOORHEAD and DANGERFIELD, 1998).

The VAE assumption is that each cash flow repeats itself perpetually, where infinity is a horizon common to all alternatives.

When using this criterion, the nature of the value of the installments must be taken into account. If they represent positive income, the project with the highest equivalent annual flow should be chosen. Otherwise, if the installments represent costs, such as annual maintenance costs, the project with the lowest equivalent annual flow in absolute terms should be chosen.

The equivalent annual value is calculated using the expression (*RESENDE and OLIVEIRA*, 2001):

$$VAE = [NPV \,.\, ((1+i) - 1) \,.\, (1+i)n] / [(1+i)n - 1]$$

Where: *EAV* = equivalent annual value; *NPV* = net present value; *i* = interest rate; *n* = project time with the longest duration, in years.

A project is considered economically viable if the EAV is positive, which indicates that the periodic benefits are greater than the periodic costs (SOARES et al, 2003).

9.5 Expected Land Value (ELV)

The expected value of the land is determined by the result of the expenses and income incurred on a unit of forest production area, discounted at a market interest rate, according to Rezende (2005), as described below.

Net income is calculated as the difference between gross income and expenses, which is a periodic income in perpetuity. It is assumed that income and expenses are equal and constant from one rotation to the next.

Since periodic income and expenses are eternal, they can be capitalized and the capitalized value is the Expected Land Value (ELV), calculated by the equation:

$$VET = \frac{R_r + \sum_{j=1}^{r} Dj * 1{,}0i^{r-j} + N_q * 1{,}0i^{r-q} - C * 1{,}0i^{r} - A * (1{,}0i^{r} - 1)}{1{,}0i^{r} - 1}$$

10 FOREST ROTATION

The expected age for the final harvest of a forest stand is called the rotation age. There are many factors that influence the rotation age, such as the price of the products obtained from the forest, costs and discount rates, future price, reinvestment options, number of rotations, replanting costs, use and price of non-timber forest products, non-marketable forest ecological services and recreational forest services. Theories developed to determine the best rotation age have used combinations of physical, biological and economic criteria.

10.1 Rotation for maximum productivity

This is the age at which the Average Annual Volume Increase (AIVI) per hectare is maximum, giving the forest stand maximum productivity. Considering an adjusted equation for the volume per hectare growth curve [V=f(t)], the age of maximum AMVI can be determined by dividing f(t) by its first derivative f '(t), using the equation t = f(t) / f '(t) - see section 3.11.4 e Figura 28 .

10.2 Rotation of maximum financial result

According to Newman (1988) there are six different main criteria for establishing the optimum age for forest rotation in economic and financial terms: maximization of gross income; maximization of annual income; maximization of net income discounted from a single rotation; maximization of net income discounted from an infinite series of similar rotations; maximization of annual net income; maximization of the growth rate of capital. However, single rotations are not common practice due to the costs of transforming forest crops into other forms of land use. In addition, in order to compare different types of enterprises with different planning horizons, infinite series must be used. Therefore, the criteria in force are for infinite series, which implies that there are only two criteria for determining the optimum rotation in financial terms: considering the cost of the land, or not considering the cost of the land. Therefore, the entrepreneur must decide which of the two criteria to use, or calculate using both and then decide. Therefore, it remains to use the criterion of the age of maximum Expected Land Value (ELV), or that of maximum Equivalent Annual Value (EAV) to determine the optimum rotation age in financial terms.

10.2.1 Maximum speed VET

The Faustmann method, commonly referred to as land expectation value, or land expected value (LEV), or soil expectation value, or soil expected value (SEV), has been frequently cited in forestry literature.

VET is a forestry term used to represent the net present value of an area of bare land to be used for timber production, calculated on the basis of an infinite series of rotations. This criterion is known worldwide and used to determine the economic rotation and maximum purchase price of bare land, considering an infinite series, as well as to select alternative projects (SILVA et al., 2002).

The VET calculation is based on perpetual net revenue (RT - CT), excluding the cost of land to be obtained from a crop (reforestation).

Since VET considers an infinite horizon, it is widely used in the economic analysis of forestry projects, as it eliminates the problem of comparing projects with different durations.

Tabela 25 - Expected Land Value (ELV) with a 21-year rotation: 3 harvests.

Ano	Discriminação	Produção (m³/ha)	Preço R$ m³	Anos*	Despesa	Despesa atualizada	Receita	Receita atualizada
Todos	Administração			21	200,00	8717,28		0,00
0	Implantação			21	3800,00	14979,91		0,00
1	Manutenção 1° ano			20	700,00	2584,97		0,00
2	Manutenção 2° ano			19	350,00	1210,76		0,00
3				18		0,00		0,00
4				17		0,00		0,00
5				16		0,00		0,00
6				15		0,00		0,00
7		240	45,00	14		0,00	10800,00	26950,95
8	Condução da brotação			13	1200,00	2805,20		0,00
9				12		0,00		0,00
10				11		0,00		0,00
11				10		0,00		0,00
12				9		0,00		0,00
13				8		0,00		0,00
14		280	45,00	7		0,00	12600,00	19904,25
15	Condução da brotação			6	1200,00	1775,78		0,00
16				5		0,00		0,00
17				4		0,00		0,00
18				3		0,00		0,00
19				2		0,00		0,00
20				1		0,00		0,00
21		220	45,00	0		0,00	9900,00	9900,00
Totais						32073,89		56755,20
Receita liquida							(valor futuro =)	24681,31
Valor esperado da terra							(valor atual =)	8389,06
VET anual								399,48

Onde: (*) Anos= anos para o final do período.

The formulas used in Tabela 25 were as follows:

- For updated administrative expenses:

$$DAa = D.((1+j)^{r-1})/j$$

Where: DAa = updated administration costs; D = costs for the year; j = interest rate; r = age of final cut.

- For expenses updated every year:

$$DA=D.(1+j)^i$$

Where: DA = updated expenditure; D = realized expenditure in the year considered; j = interest rate; i = number of years until the final cut-off.

- For updated recipes:

$$RA=R.(1+j)^i$$

Where: RA = updated revenue; R = revenue realized in the year considered; j = interest rate; i = number of years until the final cut-off.

10.2.2 Rotation of maximum VAE

Annual Equivalent Value (EAV) is the periodic and constant installment required to pay an amount equal to the NPV of the investment option under analysis, throughout its useful life. In other words, the EAV transforms the present value of the project or its NPV into a flow of periodic and continuous revenues or costs, equivalent to the present value, during the project's useful life.

An example is shown in Figura 43 a Figura 45 with rotations of 7, 8 and 9 years, respectively, at an interest rate of 6.75% p.a.

The formulas used for DA and RA are the same as those listed in the previous section for VET. The formulas for NPV and VET are as follows:

$$VAE_i = (NPV_i \,.\, ((1+j)-1)*(1+j)^i\,) \,/\, ((1+j)\,)^{i-1}$$

Where: EAV_i = equivalent annual value in year i; NPV_i = net present value in year i; i = year considered; j = interest rate.

$$NPV_i = RA - DA$$

Where: NPV = net present value; RA = sum of revenues updated up to the year in question; DA = sum of expenses updated up to the year in question; note: in this example, the updated revenue is not added because only one harvest is expected.

Ordem	Ano	Atividade	Despesa	Despesa atualizada	Soma despesa	IMA	Produção	Preço da madeira	Receita	Receita atualizada	VPL no ano	VAE
			R$	R$	R$	m³/ha/ano	m³/ha	R$/m³	R$	R$	R$	R$
1	0	Implantação	3500,00	3500,00	3500,00				0,00	0,00	-3500,00	-
2	1	Manutenção 1	700,00	655,74	4155,74				0,00	0,00	-4155,74	-4436,25
3	2	Manutenção 2	400,00	351,01	4506,75				0,00	0,00	-4506,75	-2484,01
4	3	Manutenção 3	300,00	246,61	4753,37				0,00	0,00	-4753,37	-1803,01
5	4	Manutenção	200,00	154,01	4907,38				0,00	0,00	-4907,38	-1440,63
6	5	Manutenção	200,00	144,27	5051,65	30,00	140,00	45,00	6300,00	4544,66	-507,00	-122,83
7	6	Manutenção	200,00	135,15	5186,81	31,00	180,00	45,00	8100,00	5473,66	286,85	59,72
8	7	Colheita	900,00	569,73	5756,53	31,00	231,00	52,00	12012,00	7603,97	1847,43	339,82

Figura 43 - Rotation of maximum VAE with 7-year rotation.

Ordem	Ano	Atividade	Despesa	Despesa atualizada	Soma despesa	IMA	Produção	Preço da madeira	Receita	Receita atualizada	VPL no ano	VAE
			R$	R$	R$	m³/ha/ano	m³/ha	R$/m³	R$	R$	R$	R$
1	0	Implantação	3500,00	3500,00	3500,00				0,00	0,00	-3500,00	-
2	1	Manutenção 1	700,00	655,74	4155,74				0,00	0,00	-4155,74	-4436,25
3	2	Manutenção 2	400,00	351,01	4506,75				0,00	0,00	-4506,75	-2484,01
4	3	Manutenção 3	300,00	246,61	4753,37				0,00	0,00	-4753,37	-1803,01
5	4	Manutenção	200,00	154,01	4907,38				0,00	0,00	-4907,38	-1440,63
6	5	Manutenção	200,00	144,27	5051,65	30,00	140,00	45,00	6300,00	4544,66	-507,00	-122,83
7	6	Manutenção	200,00	135,15	5186,81	31,00	180,00	45,00	8100,00	5473,66	286,85	59,72
8	7	Manutenção	200,00	126,61	5313,41	32,00	231,00	45,00	10395,00	6580,35	1266,94	233,04
9	8	Colheita	900,00	533,70	5847,11	31,00	272,00	52,00	14144,00	8387,44	2540,32	421,31

Figura 44 - Rotation of maximum VAE with 8-year rotation.

Ordem	Ano	Atividade	Despesa	Despesa atualizada	Soma despesa	IMA	Produção	Preço da madeira	Receita	Receita atualizada	VPL no ano	VAE
			R$	R$	R$	m³/ha/ano	m³/ha	R$/m³	R$	R$	R$	R$
1	0	Implantação	3500,00	3500,00	3500,00				0,00	0,00	-3500,00	-
2	1	Manutenção 1	700,00	655,74	4155,74				0,00	0,00	-4155,74	-4436,25
3	2	Manutenção 2	400,00	351,01	4506,75				0,00	0,00	-4506,75	-2484,01
4	3	Manutenção 3	300,00	246,61	4753,37				0,00	0,00	-4753,37	-1803,01
5	4	Manutenção	200,00	154,01	4907,38				0,00	0,00	-4907,38	-1440,63
6	5	Manutenção	200,00	144,27	5051,65	30,00	140,00	45,00	6300,00	4544,66	-507,00	-122,83
7	6	Manutenção	200,00	135,15	5186,81	31,00	180,00	45,00	8100,00	5473,66	286,85	59,72
8	7	Manutenção	200,00	126,61	5313,41	32,00	231,00	45,00	10395,00	6580,35	1266,94	233,04
9	8	Manutenção	200,00	118,60	5432,01	33,00	272,00	50,00	13600,00	8064,84	2632,83	436,65
10	9	Colheita	900,00	499,96	5931,97	31,00	297,00	52,00	15444,00	8579,24	2647,27	402,01

Figura 45 - Rotation of maximum VAE with 9-year rotation.

As a result, it can be seen that the highest VAE occurs in the 8-year rotation (Tabela 26), which is the rotation with the highest financial profitability using the EAV.

Tabela 26 - NPV, EAV, production and productivity from 5 to 12 years.

Rotation (years)	NPV (R$)	VAE (R$)	Production (m^3/ha)	IMA (m^3/ha/year)
5	-305,01	-73,89	140,00	28
6	665,28	138,50	180,00	30
7	1847,43	339,82	231,00	33
8	**2540,32**	**421,31**	**272,00**	**34**
9	2647,27	402,01	297,00	33
10	2377,08	334,54	310,00	31
11	2000,33	263,45	319,00	29
12	1538,00	191,07	324,00	27

11 FORESTRY PRODUCTION PLANNING

Forest production planning begins with collecting data on trees and forests, including their growth. The dendrometric characteristics of individuals, the structure of stands, growth and productive potential are essential data for production planning. Data collection allows for a description of the structure of forests, which is carried out using the dendrometric characteristics of trees and stands summarized using statistics on the data collected in forest inventories, which is then used for modelling and production forecasts with the adjusted models.

11.1 Production planning and regulation process

Production planning starts from a situation in which each plot has its production forecast for the final year of the rotation period. It should be borne in mind that what is done in one rotation has implications for the following rotation and, therefore, it is ideal to plan for a period of two rotations in order to fully regulate production.

Initial unregulated production should be calculated by adding up the production of each plot cut in each year of the planning horizon, at the end of their respective rotation periods.

One of the first steps in planning is to check whether the annual average of initial unregulated production is higher or lower than the average annual demand or consumption. If average consumption is higher than average demand, there is not enough land to sustain production at current levels. In this case, there are two ways of achieving sustainability. The first is by acquiring new areas for planting and the second is to increase productivity by improving sites or genetic improvement, or by replacing the species with a more productive one that can be used for the same purposes.

The second step in planning is to check whether the deficit years can be covered by bringing forward the cutting of areas to be cut in the following years, as long as the anticipation is not too great so as not to lower productivity excessively. Next, the remaining years of deficit should be checked and strategies drawn up for supply by purchasing wood from third parties, increasing the planted area or increasing productivity, all of which can be used at the same time.

Initially, for the example above, a period of ten years had been determined, but this was extended to two rotations, i.e. 14 years, because, as can be seen from the early cutting of a small section of plot D in the first rotation, there were implications for the production of the same plot in the second rotation.

When cutting is brought forward, as in the case of plot D, from the 7th to the 6th year, for example, the productivity per hectare according to the table goes from 230.4 m^3/ha to 194.6 m^3/ha according to the production table for the site in question. The remaining years of deficit should then be checked and strategies drawn up for supply through the purchase of wood from third parties, an increase in the planted area or an increase in productivity, with the possibility of using more than one of these resources at the same time.

Production in each year is the sum of the production of the plots that reached the rotation age in the year in question, including early cuttings.

Timber should only be sold if there is no shortage in subsequent years. In the last period, sales must take into account the shortfall in the next rotation. In the example above, all the remaining wood was sold minus an amount equal to that which had to be acquired in the 1st rotation so that there would be no shortage of wood in the next rotation.

Concepts used to regulate production:

- Consumption (C(year)) is the demand or internal consumption forecast for the year in question;
- Unregulated production (P(year)) is the production obtained by multiplying the production area by the production of the plot under consideration in the year under consideration;
- Previous balance (SA(year)) is the current balance for the previous year;
- Anticipation of cutting (PA(year)): the surplus area for the following year is determined for each plot and then the anticipated production of the surplus area is calculated with the productivity at the anticipated age;
- The current internal balance (SI(year)) is the sum of the year's production and the previous balance minus the year's consumption, whereby prepayments must have already been taken into account in the unregulated production for the year in question;
- Purchases (AQ(year)) are the volumes purchased from third parties in the year in question;

- Sales (VD(year)) are the volumes sold in the year in question;
- Current balance (S(year)) is the sum of the internal balance plus acquisitions minus sales;
- Regulated final situation (SR(year)) is the sum of unregulated production plus the previous balance plus acquisitions minus sales;
- Anticipated area (AA(plot, age): is the part of a plot's area that will be anticipated;
- A(plot) = area of the plot considered;
- P(field, age) = production of the field considered at the age considered;
- AE(field, year) = Surplus area in the field considered in the year considered;
- C(plot, year) = cut in the plot considered, necessary to supply consumption in the year considered;
- V(plot, age) = production per hectare of the plot considered at the age considered.

11.2 Example 1 - Firewood production for a dairy industry

A dairy industry uses wood in its boiler to produce hot water and steam, with an annual consumption of 4,500 m^3 of wood. The company has an area of 149 hectares of effective *Eucalyptus* plantations, managed in single coppice, with a 7-year rotation between cuts. The planning horizon is two rotations (14 years).

The calculations should be carried out as in Tabela 27 in the following steps:

1°) Production in each year is the sum of the production of the plots that reached the rotation age in the year in question, including early cuttings:

P(year) = ∑V(t, i) . A(t) , where: P(year) = Production in the year considered; V(t, i) = productivity of plot t at the cutting age considered i; and, A(t) = area of plot t;

2°) Consumption (C(year)) is the demand or internal consumption forecast for the year in question;

3°) Previous balance (SA(year)) is the current balance of the previous year (S(year-1));

4°) Anticipated cuts: the surplus area for the following year is determined for each plot and then the anticipated production of the surplus area is calculated with the productivity at the anticipated age:

AA(D,2007) = [P(D,7)-C(d,2007)] / V(D,7) ; AE(D,2007) = surplus area in plot D in 2007; C(D,2007) = cut in plot D needed to supply consumption in 2007; V(D,7)=productivity per hectare of plot D at 7 years.

AE(D,2007) = [4608-4500] / 230.4 = 0.46875 ha = AA(D,2006)

PA(D,2006) = AA(D,2006) . V(D,6) ; PA(D,2006) = anticipated production of plot D for the year 2006; AA(D,2006) = anticipated area of plot D for 2006; V(D,6) = Productivity in m^3/ha of plot D at 6 years of age;

PA(D,2006) = 0.46875 x 194.6 = 91.22 m^3 ~ 91 m^3

5°) Current internal balance (SI(year)): this is the sum of production for the year (P(year)) and the previous balance

(SA(year)) minus consumption for the year (C(year)). Note: prepayments must have already been taken into account in the unregulated production for the year in question;

6º) Purchases (AQ(year)): These are the volumes purchased from third parties in the year under consideration;

7º) Sales (VD(year)): These are the volumes sold in the year under consideration;

Tabela 27 - Flow prognosis and production regulation (m^3/ha)

Talhão	Área	h0	Produção (m^3/ha)			Ano														Totais	Médias
			V(t,7) 7 anos	V(t,6) 6 anos		2004	2005	2006	2007	2008	2009	2010	2011	2012	2013	2014	2015	2016	2017		
A	24.5	22	194.4	163.9	Idade	1	2	3	4	5	6	7	1	2	3	4	5	6	7	9526	-
					Prod.							4763							4763		
B	17.0	26	268.9	227.6	Idade	2	3	4	5	6	7	1	2	3	4	5	6	7	1	9143	-
					Prod.						4571							4571			
C	22.6	24	230.4	194.6	Idade	3	4	5	6	7	1	2	3	4	5	6	7	1	2	10414	-
					Prod.					5207							5207				
D	20.0	24	230.4	194.6	Idade	4	5	6	7	0	1	2	3	4	5	6	0	1	2	9199	-
					Prod.			91	4500						108	4500					
E	22.1	22	194.4	163.9	Idade	5	6	7	0	1	2	3	4	5	6	7	1	2	3	8592	-
					Prod.			4296							4296						
F	17.5	24	230.4	194.6	Idade	6	7	1	2	3	4	5	6	7	1	2	3	4	5	8064	-
					Prod.		4032							4032							
G	25.3	22	194.4	163.9	Idade	7	1	2	3	4	5	6	7	1	2	3	4	5	6	9837	-
					Prod.	4918							4918								
Produção não regulada ($P_{(ano)}$)						4918	4032	4387	4500	5207	4571	4763	4918	4032	4296	4500	5207	4571	4763	64667	4619
Consumo ($C_{(ano)}$)						-4500	-4500	-4500	-4500	-4500	-4500	-4500	-4500	-4500	-4500	-4500	-4500	-4500	-4500	-63000	-4500
Saldo anterior ($SA_{(ano)}$)						0	418	0	0	0	37	108	371	789	321	0	0	162	233	-	-
Antecipação de corte ($PA_{(ano)}$)						0	0	91	0	0	0	0	0	0	0	707	0	0	0	-	-
Saldo interno atual ($SI_{(ano)}$)						418	-50	-113	0	707	108	371	789	321	118	0	707	233	496	-	-
Aquisições ($AQ_{(ano)}$)						0	50	113	0	0	0	0	0	0	0	0	0	0	0	163	-
Vendas ($VD_{(ano)}$)						0	0	0	0	-670	0	0	0	0	-118	0	-545	0	-333	-1666	-
Saldo atual ($S_{(ano)}$)						**418**	**0**	**0**	**0**	**37**	**108**	**371**	**789**	**321**	**0**	**0**	**162**	**233**	**163**	-	-
Situação final regulada ($SR_{(ano)}$)						**4918**	**4500**	**4500**	**4500**	**4537**	**4608**	**4871**	**5289**	**4821**	**4500**	**4500**	**4662**	**4733**	**4663**	65604	

8th) Current balance (S(year)): this is the sum of the internal balance (SI(year)) plus acquisitions (AQ(year)) minus sales (VD(year));

9th) Regulated final situation (SR(year)): This is the sum of unregulated production (P(year)) plus the previous balance (SA(year)) plus acquisitions (AQ(year)) minus sales (VD(year));

10th) Timber sales: in this example, all the surplus timber was sold, minus an amount equal to that which had to be purchased in the 1st rotation so that there would be no shortage of timber in the next rotation.

This is a simple example, without thinning, with only one type of wood and cutting one plot each year. There are usually several plots to cut. When areas are acquired for planting, they should be included in the planning as new stands. In more complex cases, there may be high-stem management with thinning and several wood assortments. In this case, it is necessary to plan the production of each assortment separately, as well as the supply strategies for each assortment, according to the consumption forecast for each one.

A summary of the final situation is shown in Figura 46 where the negative values (sales and consumption) are separated from the positive ones (unregulated production, previous balance and acquisitions).

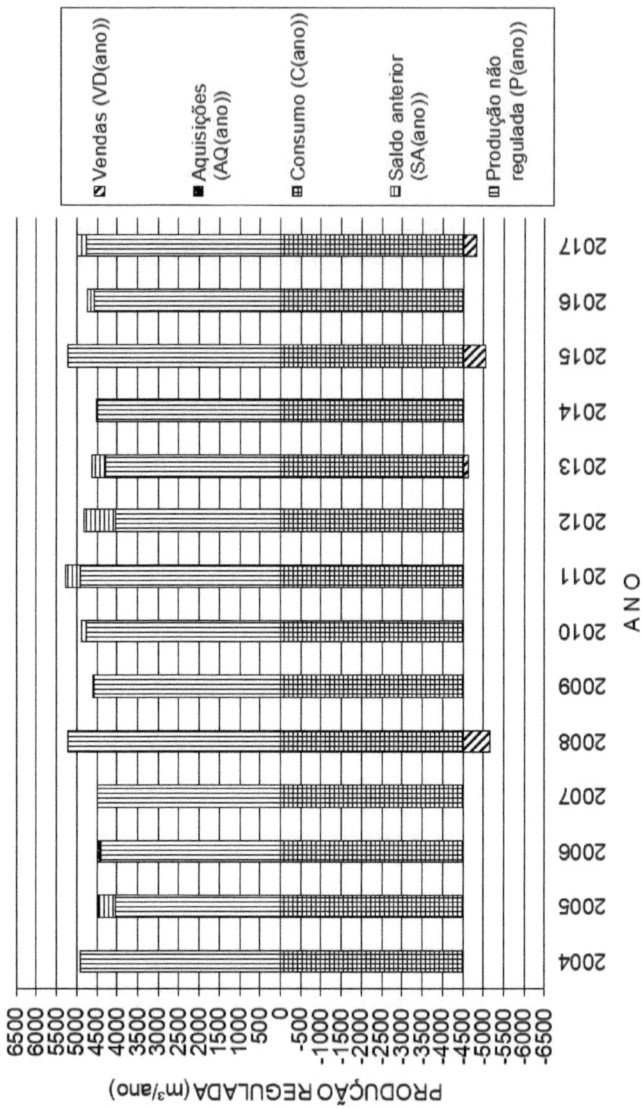

Figura 46 - Results of production regulation .

11.3 Example 2 - Regulating the production of firewood for a refrigerator

A slaughterhouse uses firewood in its boiler and has a forest production area of 1428 ha. Its wood consumption is 45,000m³/year and it is managed using simple coppice. The forestry data is shown in Tabela 28 and Tabela 29 .

Considering a rotation of 7 years and a planning horizon of 11 years, what is the production plan for a constant supply?

Tabela 28 - Production by age and Site Index (SI) in m³/ha.

Age	Site Index (SI)			
	20	22	24	26
2	20.7	25.5	30.7	36.4
3	47.5	58	69.5	82.1
4	77.5	94.3	112.6	132.4
5	107.4	130.3	155.1	181.8
6	135.4	163.9	194.6	227.6
7	161	194.4	230.4	268.9

Tabela 29 - Data on the company's forest stands obtained from the current year's Forest Inventory.

Plot	Area	Age	h_{100}	Production (m³/ha)	Stock (m³)
1	35	1	20	0.0	0.0
2	50	1	24	0.0	0.0
3	22	1	24	0.0	0.0
4	32	1	20	0.0	0.0
5	40	1	26	0.0	0.0
6	30	1	24	0.0	0.0
7	21	2	26	36.4	764.4
8	27	2	26	36.4	982.8
9	30	2	24	30.7	921.0
10	39	2	26	36.4	1419.6
11	45	2	26	36.4	1638.0
12	25	3	20	47.5	1187.5
13	49	3	26	82.1	4022.9

Plot	Area	Age	h_{100}	Production (m³/ha)	Stock (m³)
14	22	3	24	69.5	1529.0
15	37	3	26	82.1	3037.7
16	38	3	26	82.1	3119.8
17	42	3	24	69.5	2919.0
18	24	4	26	132.4	3177.6
19	26	4	20	77.5	2015.0
20	45	4	24	112.6	5067.0
21	29	4	24	112.6	3265.4
22	36	4	24	112.6	4053.6
23	38	4	26	132.4	5031.2
24	48	5	22	130.3	6254.4
25	25	5	24	155.1	3877.5
26	30	5	20	107.4	3222.0
27	35	5	22	130.3	4560.5
28	26	5	24	155.1	4032.6
29	44	5	24	155.1	6824.4
30	29	6	22	163.9	4753.1
31	20	6	22	163.9	3278.0
32	40	6	24	194.6	7784.0
33	45	6	24	194.6	8757.0
34	20	6	24	194.6	3892.0
35	29	6	24	194.6	5643.4
36	47	7	20	161.0	7567.0
37	42	7	22	194.4	8164.8
38	30	7	24	230.4	6912.0
39	50	7	22	194.4	9720.0
40	41	7	24	230.4	9446.4
41	45	7	20	161.0	7245.0
Total	1428	-	-	-	156085.6

Production planning and regulation is carried out as follows Tabela 30 in which total annual production is forecast according to the management regime adopted, taking into account production by Site Index, the necessary advances and purchases and any surpluses sold.

A graph summarizing the results can be drawn up as in Figura 47 sales from last year are not debited. It can be seen that the result of the minimum amount of wood made available is 45,000 m², which is the annual consumption forecast by the industry.

Tabela 30 - Planning and regulating production per year (m³/year).

Plot	Area	IS	Year of production										
	(ha)	(m)	1	2	3	4	5	6	7	8	9	10	11
1	35	20							5635				
2	50	24							11520				
3	22	24							5069				
4	32	20							5152				
5	40	26							10756				
6	30	24							6912				
7	21	26						5647					
8	27	26						7260					
9	30	24						6912					
10	39	26						10487					
11	45	26						12101					
12	25	20					4025						
13	49	26					13176						
14	22	24					5069						
15	37	26					9949						
16	38	26					10218						
17	42	24					9677						
18	24	26				6454							6454
19	26	20				4186							4186
20	45	24				10368							10368
21	29	24				6682							6682
22	36	24				8294							8294
23	38	26			**1030**	9016							10218
24	48	22			9331							9331	
25	25	24			5760							5760	
26	30	20			4830							4830	
27	35	22			6804							6804	
28	26	24			5990							5990	
29	44	24			10138							10138	
30	29	22		5638							5638		
31	20	22		3888							3888		
32	40	24		9216							9216		
33	45	24		10368							10368		
34	20	24		4608							4608		
35	29	24		6682							6682		
36	47	20	7567							7567			

			Year 1	Year 2	Year 3	Year 4	Year 5	Year 6	Year 7	Year 8	Year 9	Year 10	Year 11
37	42	22	8165							8165			
38	30	24	6912							6912			
39	50	22	9720							9720			
40	41	24	9446							9446			
41	45	20	7245							7245			
Total	1428		49055	40399	43883	45000	52114	42407	45044	49055	40399	42853	46202
Consumption			45000	45000	45000	45000	45000	45000	45000	45000	45000	45000	45000
Previous balance			0	4055	0	0	0	5242	2649	2693	6748	2147	0
Anticipating the cut			0	0	1030	0	0	0	0	0	0	0	0
Own balance			4055	-546	-1117	0	7114	2649	2693	6748	2147	0	1202
Acquisitions			0	546	1117	0	0	0	0	0	0	0	0
Sales							1872						1202
Current balance			**4055**	**0**	**0**	**0**	**5242**	**2649**	**2693**	**6748**	**2147**	**0**	**0**
Regulated final situation			**49055**	**45000**	**43970**	**45000**	**50242**	**47649**	**47693**	**51748**	**47147**	**45000**	**45000**

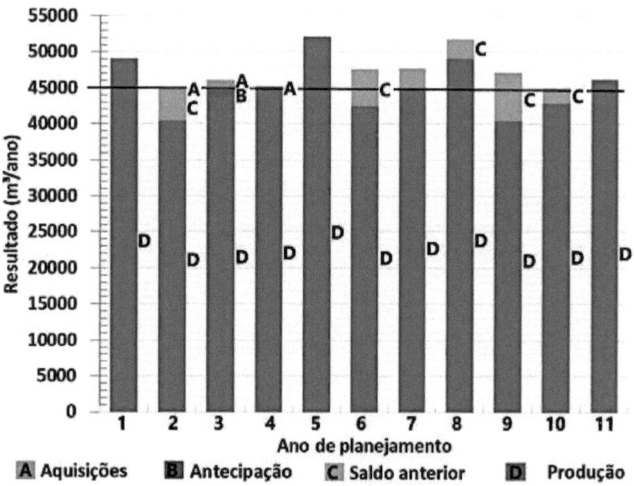

Figura 47 - Annual production after prepayments, acquisitions and sales, reaching or exceeding the minimum production (without discounting last year's sales).

12 FORESTRY OPERATIONAL RESEARCH (PO)

Operations research is an applied science that aims to optimize the use of production resources through quantitative methods. It is a tool for planning production using mathematical methods.

12.1 Introduction

The term Operations Research (OR) was coined during World War II, when the British military administration called in a group of scientists to apply scientific methods to the study of military operations with the aim of winning battles.

The main objective was to use military resources effectively for various operations and for the various activities within each operation.

Since then, the efficiency of operations research in the military has aroused the interest of other areas of government and industry.

After World War II, a number of factors contributed to the development and use of OP in other areas:

- Availability of PO scientists who have worked in the armed forces;
- Development of the Simplex method for solving linear programming problems in 1947 by George Dantzig;
- Development of computers capable of processing PO data from the 50s onwards.

Many of the OP techniques were developed before the end of the 1950s, such as:

- Linear programming;
- Dynamic programming;
- Queue theory;
- Inventory theory.

Operations Research (OR) can be understood as the use of mathematics and quantitative methods to support decision-making. OR uses tools from different areas such as mathematics, statistics, engineering, economics, psychology, among others. The PO methodology seeks to mathematically define the consequences resulting from different possible alternatives of actions to be taken to achieve a given objective. The following synonyms are used for PO: decision analysis, operations research, operations research, scientific administration, scientific management, mathematical programming.

The OP involves six main phases for its development:

1) Problem formulation
2) Model development
3) Selecting the type of input data
4) Model solution
5) Model validation
6) Implementing the solution

The problem formulation consists of:

- Defining the objective;

- Determination and quantification of decision alternatives or task execution;
- Determining and quantifying the constraints for carrying out tasks.

The following aspects must be taken into account when developing models:

- Once the problem has been defined, it needs to be transformed into a mathematical model that represents it;
- Static and dynamic elements must be identified and mathematical resources used to express the relationships between them;
- The proposed model will have to be tested in the field and modified under environmental conditions in order to satisfy all the constraints and achieve the objectives.

The selection of input data is critical to the entire process:

- No model works without appropriate data;
- The aim of this phase is to ensure that the quality and quantity of data is sufficient to test and operate the mathematical model with regard to the objective, alternatives and restrictions.

In the model's solution phase, the data is preliminarily processed taking into account that:

- In this phase, the mathematical model is operated in a test phase;
- If the model does not work properly or if the expected objectives are not achieved, consideration should be given to modifying the model or the input data.

To validate the model, it should be borne in mind that:

- A model is considered valid if it allows a real prognosis of the system's performance;
- A model must be applicable in the long term and must be updated periodically, taking into account aspects of the past, present and future.

When implementing the solution, it should be borne in mind that in this phase the best solution found by means of the mathematical model must be planned, executed, controlled, analyzed and modified when necessary. To this end, Demming's PDCA (Plan, Do, Control, Act) cycle can be used to implement the solution.

12.2 Operational Research Methods

There are many methods used in PO, among them the following:

- LINEAR PROGRAMMING - This is a mathematical method for allocating a fixed amount of resources in which certain objectives are optimized and certain constraints are met;
- INTEGRAL PROGRAMMING - This is a special case of linear programming in which one or more variables only take on integer values;
- OBJECTIVE-BASED PROGRAMMING - A methodology for dealing with multiple and incompatible organizational objectives;
- TRANSPORT LOGISTICS - This is a special type of linear programming in which the objective is to minimize the cost of distributing a product from various sources to various destinations;

- QUEUE THEORY - A method for optimizing the duration of sequential tasks;
- GAME THEORY - It is an analysis method for every decision when there are one or more opponents in a situation of conflicting objectives, in order to optimize results, maximizing profits and minimizing losses;
- INVENTORY CONTROL - It is a mathematical method for reordering time periods, levels and quantities, ensuring supply, minimizing waste and losses. It relates to the acquisition, stocking and handling of products;
- SIMULATION TECHNIQUE - This is a methodology for creating models from real situations and applying them in experiments to determine possible outcomes. It is used when the risks in a real situation are too great, or the real situations are not manageable, or the time consumed in a real situation is too great;
- ALLOCATION THEORY - This is a method that involves allocating different resources to different tasks;
- NON-LINEAR PROGRAMMING - This is an optimization methodology used when the objectives or constraints are of a non-linear nature, such as the growth of trees or stands, as well as discount rates applied to bulk purchases;
- DYNAMIC PROGRAMMING - A methodology used to solve problems that involve making decisions about a series of consecutive activities;
- SEQUENTIAL THEORY - This is a special type of queue theory in which the resources are fixed, but the order of activities can be controlled. Example: several planes waiting to land;
- REPLACEMENT MODELS - These are techniques for optimizing the replacement of resources (products, machines, personnel, capital) due to their depletion, deterioration, failure, retirement, etc;

- MARKOV CHAIN - This is a probabilistic method for determining future stage situations. It is used when a process can be separated into several stages and the probability of passing from one stage to the next is known, such as the natural regeneration of forest populations;
- MARKOV DECISION PROCESS - A process based on the Markov chain that helps choose actions to optimize results, taking into account what happened immediately before and after;
- PERT AND CPM NETWORK PLANNING - PERT (Program Evaluation and Review Technique) and CPM (Critical Path Method) are techniques used for planning and monitoring large projects. It is useful for scheduling tasks based on the mandatory precursors, or subsequent tasks, and the duration of each one;
- SYMBOLIC LOGIC - A methodology for replacing symbols with logic, used mainly in the computer design of machines and circuits;
- INFORMATION THEORY - This is an analytical process borrowed from the field of electronic communication. It seeks to evaluate the efficiency of the flow of information in a system and helps to improve its performance.

12.3 Benefits of PO

The main advantages of PO are as follows:

- Control - Provides analytical and quantitative tools to identify problems and find the best solution, with wide application in production and supply planning;
- System - Assists in decision-making for the best allocation of resources, reducing their use due to optimization, facilitating the planning of operations and identifying obsolete resources, which results in more efficient systems;

- Decision-making - Mathematical models help to reduce errors in decision-making by indicating the best course of action;
- Coordination - Operational research oriented towards planning models can help coordinate different areas of an organization.

12.4 Limitations of the OP

Among the limiting factors in OP, the following can be mentioned:

- Electronic resources - Depends on computer resources such as computers and especially specific software;
- Non-quantifiable factors - Some variables are difficult to quantify and are often quantified subjectively, which can lead to errors in the solution;
- Distance between Management and Scientific Administration - The execution of PO models may not be easy for the administrator, just as it may be difficult for the PO scientist to understand a given real system if it is not his field of work;
- Cost and time required - When a database undergoes constant change, incorporating it into the PO model can involve a substantial cost. Sometimes a moderate solution at present can be more desirable than a perfect PO solution;
- Implementation - Implementing decisions is a delicate task, involving working and human relationships that are not always very straightforward.

12.5 PO applications

Among the problems that can be solved by scientific management are the following:

1. Financial economics:
   - Credit analysis;
   - Cash flow analysis;
   - Dividend policy;
   - Investments.
2. Marketing
   - Product selection;
   - Rental of advertising resources;
   - Number of salespeople;
   - Product mix selection.
3. Sales, acquisition and exploitation
   - Optimization of purchase orders;
   - Replacement policy.
4. Production management
   - Location and sizing of resources;
   - Distribution policy;
   - Loading and unloading resources;
   - Production planning;
   - Project management.
5. Personnel management
   - Personnel selection;
   - Recruitment;
   - Transfers;
   - Allocation and development.
6. Research and development
   - Project selection;
   - Project control;
   - Consistency and choice of alternatives.

12.6 Linear Programming (PL)

PL is a mathematical methodology for choosing alternatives for the use of different resources in order to optimize objectives under certain constraints. The objective of PL can be to maximize or minimize results.

The terminology used in PL is different from other disciplines, so it is necessary to know its concepts. The main terms and expressions are described below:

Linear function - a linear function is composed of additive terms in which each variable is related to only one coefficient and each coefficient is related to only one variable, of the type z=f(x), where z is unknown and dependent on x; x = $\sum$ cj . xj, where xj is known, independent and xj $\geq$ 0, called decision variables; cj are coefficients to be determined, called decision parameters; j = order number of the variable (=1 to n).

Objective function - is a linear function of decision variables (x_j) that expresses the objective of the decision as follows: z = f(x); objective functions seek to maximize f(x) or minimize f(x), the objective z.

Decision variables - are the factors (x_j) that influence the decision to optimize the objective function z; the variables can be controlled by the programmer in values to be determined and whose quantities (c_j) of use of each one (coefficients of use, or decision parameters) are the unknowns sought to optimize the objective.

Constraints - are influencing factors on the decision variables that limit their use, expressed mathematically as: $f(x) \geq b_i$, $f(x) \leq b_i$, or $f(x) = b_i$, where: $x = \sum a_i \cdot x_j$; b_i = value of constraint i, called the right hand side (RHS); i = order number of the constraint (= 1 to m).

Admissible solution - is any non-negative solution that satisfies the constraints; the space comprising all admissible solutions is called the admissibility region.

Optimal solution - is the admissible solution that optimizes (maximizes or minimizes) the objective z.

The general mathematical model of PL is based on a system of equations and inequalities, as follows:

Optimize (maximize or minimize) z by the following equation:

$$z = c1x1 + c2x2 + \ldots + cjxj$$

Subject to:

$$a11x1 + a12x2 + \ldots + a1nxn \ (\geq,=,\leq) \ b1$$
$$a21x1 + a22x2 + \ldots + a2nxn \ (\geq,=,\leq) \ b2$$
$$\ldots \quad \ldots \quad \ldots \quad \ldots \quad \ldots \quad \ldots$$
$$am1x1 + am2x2 + \ldots + amnxn \ (\geq,=,\leq) \ bm$$

$$x1, x2, \ldots, xn \geq 0$$

There is a plethora of software, including free software for solving mathematical programming problems, including LibreOffice-CALC, EXCEL-SOLVER, LINGO and SAS. LibreOffice-CALC is free software, and LINGO comes in a free academic version and a paid version, while EXCEL-SOLVER and SAS are proprietary software.

12.6.1 Linear programming models in forest management

The application of linear programming in forestry began in the 1960s (CURTIS, 1962; LEAK, 1964; LOUCKS, 1964; NAUTIYAL AND PEARSE, 1967). Forest planning by linear programming began in 1971 with Navon and the development of the Timber RAM (FS-USA) and with Ware and Clutter at the University of Georgia. The "*Timber Resource Allocation Method*" (Timber RAM) gave rise to the FORPLAN systems (IVERSON AND ALSTON, 1986) and recently SPECTRUM (USA, 1999), both from FS-USA.

In Brazil, the first applications published in the forestry area were by Rodriguez and Lima (1985) and Rodriguez et al (1986).

Johnson and Scheurman (1977), analyzing how variables are defined in these applications, defined two basic structures, which they called Model I and Model II. From the basic model called Model I, many others were generated to try to remedy the deficiencies found, the best known being Model II. Models I and II do not define the spatial location of the management units, which is why they are used in the strategic planning of forest resource management.

Most other LP models are derivations of the first two, or combinations of both. In tactical and operational planning, other models and other mathematical programming techniques are used, such as Integer Programming (IP),

Mixed Integer Programming (MIP), Non-Linear Programming (NLP), Multiobjective Programming (MOP), Critical Path Method (CPM), among others. Currently, Models I and II are still very useful, but new techniques have been added to linear programming and, in many cases, are more appropriate or complementary.

DIFFERENCES BETWEEN MODELS I AND II

Model I: is more rigid; preserves the Management Units (Districts, Plots, Sections) from the beginning to the end of the planning horizon; generally has a greater number of decision variables; requires explicit variables in the final stock.

Model II: is more flexible; does not preserve Management Units during the planning period (MUs can be divided or merged during the planning period); has a greater number of area restrictions; the volume of felling is generally greater at the end of the planning period.

MODEL I

Objective:

$$Maximizar\ Z = \sum_{i=-1}^{N} \left(\sum_{k=1}^{M} D_{ik} X_{ik} \right)$$

Subject to:

- Area restrictions;
- Productivity restrictions;
- Operational capacity restrictions;
- Cash flow restrictions;
- Non-negativity restrictions.

Where: Z = total value of the project (value of the objective function); N = number of plots or basic management units (BMU); M = number of management regimes (prescriptions, or alternatives); X_{ik} = Area of basic management unit i allocated to prescription k; D_{ik} = Value of prescription *k* per unit area if implemented in plot i .

Variables and Parameters of the PL Basic Forestry Model I

- Z = total value of the project (value of the objective function);
- N = number of plots or basic management units (BMU);
- M = number of management regimes (prescriptions, or alternatives);
- T = number of planning years (planning horizon);
- P = number of forest products (outputs);
- A_i = Total area of basic management unit i;
- X_{ik} = Area of basic management unit i assigned to prescription k;
- D_{ik} =Value of prescription k per unit area if implemented in plot i;
- V_{iktp} = Volume (output) per unit area of product p in field i in year t obtained with prescription k;
- F_{ikt} = Inflow (+) or outflow (-) in the cash flow of plot i in year t with prescription k;
- $VMin_{tp}$ and $VMax_{tp}$ = Minimum and maximum permitted volumes of product p in planning year t;
- $OMin_t$ and $OMax_t$ = Minimum and maximum operational capacity (e.g. harvesting, planting, etc.) in planning year t;
- $FMin_t$, $FMax_t$ = Minimum and maximum cash available in year t;

- X_{ik} are the decision variables and can only take positive values. Where: i = 1,2, ... , N; k = 1,2, ... , M; t = 1,2, ... , T; p = 1,2, ... , P.

Area restrictions - refers to the physical space available for forestry production in hectares per management unit:

$$\sum_{k=1}^{M} X_{ik} \leq A_i \quad (i = 1,2,...N)$$

Where: A_i = Total area of basic management unit i; X_{ik} = Area of basic management unit i assigned to prescription k.

Productivity restrictions - refers to the stock expected at each age of each management unit in relation to the minimum and maximum consumption allowed:

$$\sum_{i=1}^{N}(\sum_{k=1}^{M} V_{iktp} X_{ik}) \geq VMin_{tp} \quad (t = 1,2,\ldots,T)\ (p = 1,2,\ldots,P)$$
$$\sum_{i=1}^{N}(\sum_{k=1}^{M} V_{iktp} X_{ik}) \leq VMax_{tp} \quad (t = 1,2,\ldots,T)\ (p = 1,2,\ldots,P)$$

Where: X_{ik} = Area of basic management unit i assigned to prescription k ; V_{iktp} = Volume (output) per unit area of product p in plot i in year t obtained with prescription k ; $VMin_{tp}$ and $VMax_{tp}$ = Minimum and maximum permitted volumes of product p in planning year t

Operational capacity restrictions - These are the operations recommended for each unit in each period:

$$\sum_{i=1}^{N}(\sum_{k=1}^{M} \alpha_{ikt} X_{ik}) \geq OMin_t \quad (t = 1,2,\ldots,T)$$
$$\sum_{i=1}^{N}(\sum_{k=1}^{M} \alpha_{ikt} X_{ik}) \leq OMax_t \quad (t = 1,2,\ldots,T)$$

Where: X_{ik} = Area of basic management unit i assigned to prescription k; $OMin_t$ and $OMax_t$ = Minimum and maximum operational capacity (e.g. harvesting, planting, etc.) in planning year t; α_{ikt} = 0 if prescription k does not recommend operating on unit i in period t; α_{ikt} =1 if prescription k recommends operating on unit i in period t.

Cash flow restrictions - These are the restrictions on the availability of financial resources in each period considered:

$$\sum_{i=1}^{N}\left(\sum_{k=1}^{M} F_{ikt} X_{ik}\right) \geq FMin_t \quad (t = 1,2,\ldots,T)$$

$$\sum_{i=1}^{N}\left(\sum_{k=1}^{M} F_{ikt} X_{ik}\right) \leq FMax_t \quad (t = 1,2,\ldots,T)$$

Where: X_{ik} = Area of basic management unit i assigned to prescription k; F_{ikt} = Inflow (+) or outflow (-) in the cash flow of plot i in year t with prescription k; $FMin_t$, $FMax_t$ = Minimum and maximum cash available in year t.

MODEL II

Objective:

$$Maximizar\ Z = \overbrace{\sum_{j=0}^{H-1} \sum_{i=-M}^{J-N} C_{ij} x_{ij}}^{A} + \overbrace{\sum_{i=-M}^{H-1} D_{iH} W_{iH}}^{B}$$

Subject to:

- Area restrictions;
- Constant flow restrictions;
- Final stock restriction;
- Non-negativity restrictions.

Where: Z = Total production = objective function; C_{ij} = Annual cost per hectare, of forest stratum i, if management regime j is employed; X_{ij} = Regeneration of areas (hectares) at the start of period i and cut down at the start of period j (followed by immediate planting); W_{iH} = Areas (hectares) regenerated at the start of period i and left for final inventory in period h, at the end of the planning horizon; H = planning horizon.

Area restriction - Ensures that the areas harvested from each age class present in period 0, plus the areas left over for the final inventory of this age class, is equal to the total area of the age class at the start of the planning period:

$$\sum_{j=0}^{H-1} x_{ij} + w_{iH} = A_i \qquad i = -M, -M+1, \ldots, -1$$

$$\sum_{k=j+N}^{H-1} x_{jk} + w_{jH} = \sum_{i=-M}^{j-N} x_{ij} \qquad j = 0, 1, 2, \ldots, H-1$$

Where: X_{ij} = Regeneration of areas (hectares) at the start of period i and cut at the start of period j (followed by immediate planting); W_{iH} = Areas (hectares) regenerated at the start of period i and left for final inventory in period h, at the end of the planning horizon; A_i = number of hectares present at the start of period 0 that were regenerated in period i, where i = -M to -1 by the progression (-M, -M+1, ..., -2, 1) A-8 represents the number of hectares that were regenerated eight periods before the start of period 0); H = planning horizon; N = minimum number of periods between cuts.

Constant flow restriction - Forces a constant flow to guarantee supply, keeping production between a minimum and a maximum:

$$(1-\alpha)F_k - F_{k+1} \leq 0 \qquad k = 0, 1, 2, \ldots, H-2$$

Where:

$$F_k = \sum_{i=-M}^{k-N} V_{ik} x_{ik} \quad F_{k+1} = \sum_{i=-M}^{(k+1)-N} V_{i(k+1)} x_{i(k+1)}$$

α = maximum reduced fraction at the cutting level allowed from period k to period k+1; β= maximum increase in fraction at the cutting level allowed from period k to period k+1: V_{ik} = volume per hectare cut at the start of period k from regeneration at the start of period i; X_{ik} = Regeneration of areas, in hectares, at the start of period i and cut at the start of period j; H = planning horizon; N = minimum number of periods between cuts.

Closing stock restriction - Guarantees a minimum stock at the end of the planning period:

$$\sum_{i=-M}^{H-N} V_{iH}^{m} W_{iH} \geq E^{m}$$

Where: V_{iH}^{m} = commercial volume of wood from the areas reforested at the beginning of period i and left in the final inventory; E^{m} = minimum commercial volume in the final inventory; W_{iH} = Areas (hectares) planted at the beginning of period i and left for the final inventory in period h, at the end of the planning horizon; The amount of inventoried commercial volume left in each age class i, at the end of the planning horizon, is given by:

$$V_{iH}^{m} W_{iH} \geq E_{i}^{m} \qquad (i = -M, -M+1, \ldots H\text{-}N)$$

E_{i}^{m} = minimum commercial volume in the final inventory in the regenerated areas at the start of period i; -M = number of periods before the start of period 0 in which the oldest class present at the start of period 0 is reforested (as an example: -M = -8 indicates that the oldest age class present was reforested eight periods before period 0); H = planning horizon; N = minimum number of periods between cuts.

Non-negativity restrictions - X and W represent area and cannot be negative:

$$X_{ij} \geq 0 \qquad para\ todas\ i\ e\ j,$$

$$W_{iH} \geq 0 \qquad para\ todas\ i$$

Where: X_{ij} = regeneration of areas (hectares) at the start of period i and cut down at the start of period j (followed by immediate planting); W_{iH} = Areas (hectares) regenerated at the start of period i and left for the final inventory in period h, at the end of the planning horizon.

12.6.2 EXAMPLE 1 - OPTIMIZING PRODUCTS TO BE MANUFACTURED

A company that manufactures 2 products wants to maximize its profit, with the profit on P1 and P2 being $50 and $60, respectively. Both require different numbers of hours from 3 machines to manufacture, as shown in Tabela 31 . The problem is to determine the combination of products that maximizes profit without violating the constraints.

Tabela 31 - Spreadsheet of machine productivity and availability data

Machine	**Product**		**Availability**
	P1	P2	(hours per week)
1	2	1	300
2	3	4	509
3	4	7	812
Profit	R$ 50	R$ 60	.

DEFINING THE OBJECTIVE:

Maximize profit.

DEFINITION OF DECISION VARIABLES:

Product P1 (x1) and product P2 (x2).

ELABORATION OF THE OBJECTIVE FUNCTION:

$$\textit{Maximize profit} = 50 \cdot x1 + 60 \cdot x2$$

DEFINITION OF RESTRICTIONS:

Machine 1: (2 . x1 + 1 . x2) ≤ 300;

Machine 2: (3 . x1 + 4 . x2) ≤ 509;

Machine 3: (4 . x1 + 7 . x2) ≤ 812;

where x1 and x2 ≥ 0.

The constraints with "e" are added together on the same line. The values to be assumed by X1 and X2 will represent the quantities of each product to be produced in order to maximize the profit to be obtained. The mathematical model is defined as an objective function and three constraints, as follows:

Objective function:

$$50.x1 + 60.x2 = maximum$$

Subject to:

$$2.x1 + 1.x2 \leq 300$$

$$3.x1 + 4.x2 \leq 509$$

$$4.x1 + 7.x2 \leq 812$$

Where: x1 and x2 ≥ 0.

Using SAS to solve the problem, you can have a program as follows:

```
* SAS PROGRAM - PL, EXAMPLE 1;
DATE A;
INPUT NAME$ X1-X2 TYPE$ RIGHT_SIDE;
DATALINES;
PROFIT 50 60 MAX .
MAQU1 2 1 LE 300
MAQU2 3 4 LE 509
MAQU3 4 7 LE 812
;
PROC LP;
ID NAME;
```

```
TYPE;
RHS RIGHT_SIDE;
RUN;
QUIT;
```

12.6.3 Example 2 - Optimizing transport to the factory

Statement:

A company has 3 forest stands with areas of 20000, 26667 and 100000 hectares and two industrial units. The company works on a cold system and the wood is stored for 3 months. Mill 1 consumes a maximum of 2 million and a minimum of 1.8 million m³ of wood per year. Mill 2 consumes a maximum of 2.4 million and a minimum of 2.2 million m³ of wood per year. The company has two intermediate warehouses with capacities of 600,000 m³ of wood each (Figura 48). The distances between the plantations, depots and mills are broken down in km in Table 32. Tabela 32 .

Tabela 32 - Distance between forest stands and deposits (Km)

Deposit	Forest stand			Factory	
	1	2	3	1	2
1	5	4	-	9	12
2	-	5	2	-	9

Note - There is no connection from Settlement 1 to Depot 2, nor from Settlement 3 to Depot 1. There is no connection from Depot 1 to Factory 2.

Problem:

Considering:

- Depot 1 capacity = 2400 thousand m³ per year;
- Depot 2 capacity = 2400 thousand m³ per year;
- Average annual forest productivity of 30 m³/ha;
- Average load capacity per truck of 40 m³/trip.

Answer:

- Which supply program minimizes transport distance?
- What is the average distance per trip?
- How many kilometers will it take to supply the two factories?
- Are there bottlenecks in production?

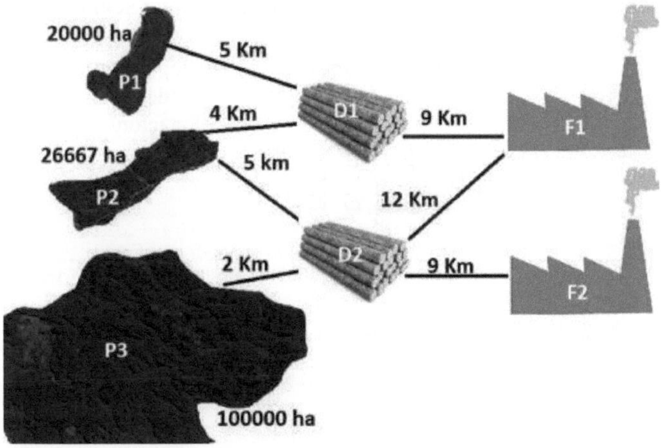

Figura 48 - Location and distances of handling units, warehouses and factories.

For the SOLVER solution, the objective is the RHS in the objective line of the spreadsheet, the decision variables are in the activity line and the constraints are the RHS values of the constraint lines.

ANALYSIS RESULTS:

- Objective function: 1160000Km will be traveled to transport the wood to the factory; Factory 1 should receive 2,000,000 m³/year and Factory 2 will receive 1,800,000 m³/year;
- The average distance traveled per trip is 12.21053 km/trip;

- Maximum annual production is: 3.800.000m³;
- Factory 2's minimum consumption will not be reached;
- The bottleneck limiting supply to Plant 2 is Depot 2, which should have the capacity to store an additional 200,000 m³ per year, or 50,000 m³ every 3 months.

Tabela 33 - Tableau corresponding to the problem for solution with EXCEL's SOLVER, or CALC.

		Decision alternatives (decision variables)						P=population; D=depot; F=factory.				
		P1.D1.F1	P2.D1.F1	P2.D2.F1	P2.D2.F2	P3.D2.F1	P3.D2.F2					
	Activity	600	800	0	0	400	2000	LHS	Relationship	RHS	Time off	Excess
	Objective	14	13	17	14	14	11		Minimum	46400		
Restrictions	Deposit 1	1	1	0	0	0	0	1400	<=	2400	1000	
	Deposit 2	0	0	1	1	1	1	2400	<=	2400	0	
	Settlement 1	1	0	0	0	0	0	600	<=	600	0	
	Settlement 2	0	1	1	1	0	0	800	<=	800	0	
	Settlement 3	0	0	0	0	1	1	2400	<=	3000	600	
	Factory 1 max	1	1	1	0	1	0	1800	<=	2000	200	
	Factory 1 min	1	1	1	0	1	0	1800	>=	1800		0
	Factory 2 max	0	0	0	1	0	1	2000	<=	2400	400	
	Factory 2 min	0	0	0	1	0	1	2000	>=	2200		-200

12.6.4 Example 3 - Optimizing harvesting machines

A company has a series of harvesting machines that allow it to assemble three types of modules to harvest the wood. The company has two objectives: maximize production and minimize costs. The machines with their costs and the modules with their possible yields are shown in Tabela 34 .

Tabela 34 - Machines, modules, costs and productivity in harvesting.

Problem data	Feller	Skidder	Processor	Forwarder	Harvester
Module1 (m^3/h)	50	15	15	0	0
Module2 (m^3/h)	0	0	0	30	12
Module3 (m^3/h)	50	0	15	30	0
Machines available	5	8	4	6	2
Cost (R$/hour)	500	400	700	500	1000

The solution involves processing in three phases. In the first, the productivity per hour must be maximized; in the second, the cost must be minimized from a minimum production to be established based on the results of the first phase, and then the data must be processed again to maximize production with a maximum cost established based on the results of the previous two phases. The table for the first phase is shown in Tabela 35 .

Tabela 35 - Maximizing productivity per hour (Tableau)

Restrictions	Module 1	Module 2	Module 3	Type	RHS
Feller	1	0	1	<=	250
Skidder	1	0	0	<=	120
Processor	1	0	1	<=	60
Forwarder	0	1	1	<=	180
Harvester	0	1	0	<=	24
Production	1	1	1	maximum	.

The SAS program to solve the problem in the first phase is as follows:

```
DATA DATA;
INPUT RESTRICTION$ MODULE_1-MODULE_3 TYPE$
RHS;
CARDS;
FELLER 1 0 1 <= 250
SKIDDER 1 0 0 <= 120
PROCESSOR 1 0 1 <= 60
FORWARDER 0 1 1 <= 180
HARVESTER 0 1 0 <= 24
PRODUCTION 1 1 1 MAX .
;
PROC LP DATA=DATA;
ID RESTRICTION;
TYPE TYPE;
RHS RHS;
RUN;
```

As a result, the objective function for maximizing productivity is 84 m³/hour. By setting a minimum of 70 m³/h as a restriction in the second phase, the table for the solution is shown in Tabela 36 . Module 1 should be used with a productivity of 60 m³/h and module 2 with a productivity of 24 m³/h. Alternatively, module 2 can be replaced by module 3 with the same productivity. 60 hours of Feller-buncher, 60 hours of Skidder, 60 hours of Processor, 24 hours of Forwarder and 24 hours of Harvester will be used.

Tabela 36 - Minimizing harvesting costs with minimum productivity of 70m³/h (Tableau)

Restrictions	Module 1	Module 2	Module 3	Type	RHS
Feller	1	0	1	<=	250
Skidder	1	0	0	<=	120
Processor	1	0	1	<=	60
Forwarder	0	1	1	<=	180
Harvester	0	1	0	<=	24
Production	1	1	1	>=	70
Cost	1600	1500	1700	minimum	.

12.6.5 Example 4 - Forest production planning

A forester has a small forest with four plots with the characteristics of Tabela 37 . The forest's continuous forest inventory data is presented in Tabela 38 .

Tabela 37 - Forest plots on the property .

Plot	Age	Area (ha)
1	2	20
2	7	25
3	14	28
4	20	23

The forester wants to minimize costs with a minimum production of 90% of the forest's productive capacity. To make things easier, an equation was used to estimate annual production, as shown in Tabela 38 .

Tabela 38 - Plots and annual stock by age and regression variables (Y, X1, x2, x3) for adjusting the equation $V=b+b_{01}.t+b_2.t^2+b_3.t^3$.

Plot	t=Age (years)	Production (m^3/ha)	Productivity ($m^3.ha^{-1}.year)^{-1}$	Y	X1=t	X2=t^2	X3=t^3
1	5	75	15.0	75	5	25	125
1	10	150	15.0	150	10	100	1000
1	15	350	23.3	350	15	225	3375
1	18	500	27.8	500	18	324	5832
1	20	600	30.0	600	20	400	8000
1	22	650	29.5	650	22	484	10648
2	5	78	15.6	78	5	25	125
2	10	154	15.4	154	10	100	1000
2	15	353	23.5	353	15	225	3375
2	18	494	27.4	494	18	324	5832
2	20	596	29.8	596	20	400	8000
2	22	653	29.7	653	22	484	10648
3	5	78	15.6	78	5	25	125
3	10	144	14.4	144	10	100	1000
3	15	343	22.9	343	15	225	3375
3	18	510	28.3	510	18	324	5832
3	20	620	31.0	620	20	400	8000
3	22	657	29.9	657	22	484	10648
4	5	71	14.2	71	5	25	125

4	10	150	15.0	150	10	100	1000
4	15	345	23.0	345	15	225	3375
4	18	517	28.7	517	18	324	5832
4	20	619	31.0	619	20	400	8000
4	22	659	30.0	659	22	484	10648

Results of the regression analysis; B0=296.8017271; B1=-81.3537999; B2=8.375974905; B3=-0.178303564; R^2=0.9975; CV=3.07%.

The graph produced with the regression equation to estimate production per hectare is outlined in Figura 49 .

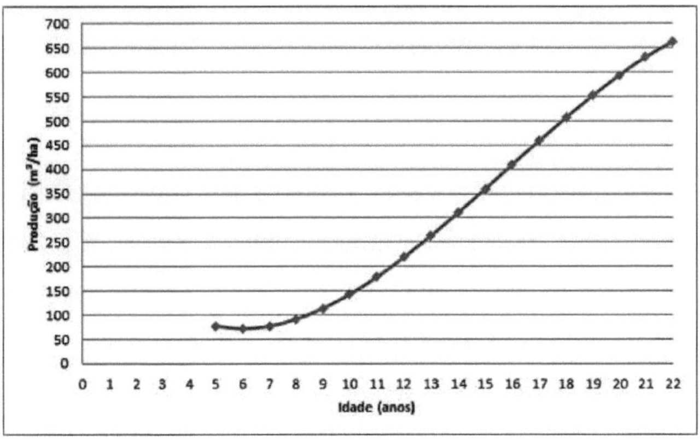

Figura 49 - Production (stock) per hectare by age (m³/ha).

The rotation ages were selected according to the dimensions of the wood desired, with a minimum of 18 years having been established, which could alternatively be 20 or 22 years (Tabela 39).

Tabela 39 - Annual production and productivity by rotation age.

Age	Production (V=m³/ha)	Productivity (m³/ha/year)
5	77	15.4
6	72	12.0

Age	Production (V=m³/ha)	Productivity (m³/ha/year)
7	77	10.9
8	91	11.3
9	113	12.6
10	143	14.3
11	178	16.2
12	219	18.2
13	263	20.2
14	310	22.2
15	359	24.0
16	409	25.6
17	458	27.0
18	506	28.1
19	552	29.0
20	594	29.7
21	631	30.0
22	662	30.1

Based on the ages of the plots and the rotation ages chosen, the possible management regimes (R) were established for each plot and the annual yields per plot in each management regime (T1R1 to T4R4) (Tabela 40).

With the management regimes established for the plots and their respective yields, production was optimized, disregarding costs, in order to obtain a value for maximum total production (Tabela 41), resulting in a maximum of 4414 m³, using 20 ha of plot 1 in management regime 3 (one rotation of 22 years), 24 ha of plot 2 in management regime 3 (one rotation of 22 years), 28 ha of plot 3 in management regime 5 (rotations of 20 and 22 years) and 23 ha of plot 4 in management regime 3 (two rotations of 22 years).

the costs per m³ produced at the different rotation ages as follows:

- 18 years: 35 R$/m³;
- 20 years: 30 R$/m³;
- 22 years: 28 R$/m³.

Using the annual production costs, the table was drawn up to minimize costs, with a minimum production of 4000 m³ in the period, as shown in Tabela 43 resulting in a minimum total cost of R$ 3770.20, which must be used:

- 20 ha of plot 1 in the management regime (a rotation of 22 years);
- 25 ha of plot 2 under management regime 3 (a rotation of 22 years);
- 21.1 ha of plot 3 under management regime 5 (a 20-year rotation and a 22-year rotation);
- 23 ha of plot 4 under management regime 3 (two 22-year rotations).

Tabela 40 - Prognosis of annual yields per plot in each possible management regime for the period from 2018 to 2045 per period.

Talhão			1	1	1	2	2	2	3	3	3	3	3	4	4	4
Regime de Manejo	Alternativas	→	T1R1	T1R2	T1R3	T2R1	T2R2	T2R3	T3R1	T3R2	T3R3	T3R4	T3R5	T4R1	T4R2	T4R3
Rotações	(anos)		18	20	22	18	20	22	18;18	18;20	18;22	20;20	20;22	22;18	22;20	22;22
Produtividade ($m^3.ha^{-1}.ano^{-1}$)	18anos		28.1	28.1	28.1	28.1	28.1	28.1	28.1	28.1	28.1	28.1	28.1	28.1	28.1	28.1
	20anos		29.7	29.7	29.7	29.7	29.7	29.7	29.7	29.7	29.7	29.7	29.7	29.7	29.7	29.7
	22anos		30.1	30.1	30.1	30.1	30.1	30.1	30.1	30.1	30.1	30.1	30.1	30.1	30.1	30.1
Produtividade por ano de produção (m³/ha)	2018	Período 1														
	2019													30.1	30.1	30.1
	2020															
	2021								28.1	28.1	28.1					
	2022															
	2023											29.7	29.7			
	2024															
	2025	Período 2														
	2026															
	2027															
	2028					28.1										
	2029															
	2030						29.7									
	2031															
	2032	Período 3						30.1								
	2033		28.1													
	2034															
	2035			29.7												
	2036															
	2037				30.1									28.1		
	2038															
	2039	Período 4							28.1						29.7	
	2040															
	2041									29.7						30.1
	2042															
	2043										30.1	29.7				
	2044															
	2045												30.1			
	Total		28.1	29.7	30.1	28.1	29.7	30.1	56.3	57.8	58.2	59.4	59.8	58.2	59.8	60.2

Tabela 41 - Tableau for maximizing production with EXCEL

Restrições	Variáveis (=Alternativas de manejo de cada talhão)														Atividade	Relação	RHS
	T1R1	T1R2	T1R3	T2R1	T2R2	T2R3	T3R1	T3R2	T3R3	T3R4	T3R5	T4R1	T4R2	T4R3			
Atividade =	0	0	20	0	0	25	0	0	0	0	28	0	0	23			
Objetivo	28.1	29.7	30.1	28.1	29.7	30.1	56.3	57.8	58.2	59.4	59.8	58.2	59.8	60.2		Máximo	4414.2
Area T1	1	1	1	0	0	0	0	0	0	0	0	0	0	0	20	<=	20
Area T2	0	0	0	1	1	1	0	0	0	0	0	0	0	0	25	<=	25
Area T3	0	0	0	0	0	0	1	1	1	1	1	0	0	0	28	<=	28
Area T4	0	0	0	0	0	0	0	0	0	0	0	1	1	1	23	<=	23

The cost per m³ produced in each regime of each plot per year of production was then determined in Tabela 42 considering the costs per m³ produced at the different rotation ages as follows:

- 18 years: 35 R$/m³;

- 20 years: 30 R$/m³;
- 22 years: 28 R$/m³.

Using the annual production costs, the table was drawn up to minimize costs, with a minimum production of 4000 m³ in the period, as shown in Tabela 43 resulting in a minimum total cost of R$ 3770.20, which must be used:

- 20 ha of plot 1 in the management regime (a rotation of 22 years);
- 25 ha of plot 2 under management regime 3 (a rotation of 22 years);
- 21.1 ha of plot 3 under management regime 5 (a 20-year rotation and a 22-year rotation);
- 23 ha of plot 4 under management regime 3 (two 22-year rotations).

Tabela 42 - Costs per m³ produced per year of production for each plot.

Talhão	RM	Rotações	Custos (R$.ha^{-1}.ano^{-1})			Anos da produção																												Total
	(Alternativas)	(anos)	18anos	20anos	22anos	2018	2019	2020	2021	2022	2023	2024	2025	2026	2027	2028	2029	2030	2031	2032	2033	2034	2035	2036	2037	2038	2039	2040	2041	2042	2043	2044	2045	
1	T1R1	18	35	30	28																35													35
1	T1R2	20	35	30	28																		30											30
1	T1R3	22	35	30	28																				28									28
2	T2R1	18	35	30	28											35																		35
2	T2R2	20	35	30	28													30																30
2	T2R3	22	35	30	28															28														28
3	T3R1	18;18	35	30	28				35																		35							70
3	T3R2	18;20	35	30	28				35																				30					65
3	T3R3	18;22	35	30	28				35																						28			63
3	T3R4	20;20	35	30	28						30																				30			60
3	T3R5	20;22	35	30	28						30																						28	58
4	T4R1	22;18	35	30	28		28																		35									63
4	T4R2	22;20	35	30	28		28																				30							58
4	T4R3	22;22	35	30	28		28																						28					56

Tabela 43 - Tableau for cost minimization with EXCEL.

Restrições	Variáveis (=Alternativas de manejo de cada talhão)														Atividade	Relação	RHS
	T1R1	T1R2	T1R3	T2R1	T2R2	T2R3	T3R1	T3R2	T3R3	T3R4	T3R5	T4R1	T4R2	T4R3			
Atividade =	0	0	20	0	0	25	0	0	0	0	21.1	0	0	23			
Objetivo	35.0	30.0	28.0	35.0	30.0	28.0	70.0	65.0	63.0	60.0	58.0	63.0	58.0	56.0		Minimo	3770.2
Produção	28.1	29.7	30.1	28.1	29.7	30.1	56.3	57.8	58.2	59.4	59.8	58.2	59.8	60.2	4000.0	=	4000.0
Area T1	1.0	1.0	1.0	0.0	0.0	0.0	0.0	0.0	0.0	0.0	0.0	0.0	0.0	0.0	20.0	<=	20.0
Area T2	0.0	0.0	0.0	1.0	1.0	1.0	0.0	0.0	0.0	0.0	0.0	0.0	0.0	0.0	25.0	<=	25.0
Area T3	0.0	0.0	0.0	0.0	0.0	0.0	1.0	1.0	1.0	1.0	1.0	0.0	0.0	0.0	21.1	<=	28.0
Area T4	0.0	0.0	0.0	0.0	0.0	0.0	0.0	0.0	0.0	0.0	0.0	1.0	1.0	1.0	23.0	<=	23.0

12.7 CPM - Critical Path Method

The critical path method (CPM) is a mathematical method for organizing one or more sequences of activities in a project into a minimum execution time.

Scheduling can be done forwards, from a start date, or backwards, from an end date.

12.7.1 Forward Programming (Early Date)

Forward scheduling starts from the source event and determines the PDIi (earliest start date of event i), which represents the longest path between the project source and event i.

$$PDIi = \max C\ (origin, i)$$

12.7.2 Backward Programming (Late Date)

Backward scheduling starts from the reverse path, i.e. from the project's objective event to determine the UDAi (Last date to finish an event i), assuming that PDIobjective = UDAobjective minus the longest path between event i and the objective event.

$$UDAi = UDAobjective - \max C\ (i, objective)$$

12.7.3 Determining the Critical Path:

The sequence of calculation steps is carried out as follows:

PDA - First Date to Finish:

$$PDAij = PDIij + dij;$$

UDI - Last Start Date:

$$UDIij = UDAij- dij;$$

FT - Total slack:

$$FTij = UDAij - (PDIij + dij);$$

FL - Free time off:

$$FLij = PDIjk - (PDIij + dij);$$

Activity ij belongs to the critical path if the early date is the same as the late date, i.e. PDI = UDA in events i and j, and there are no gaps.

Example CPM programming problem

Consider the schedule of 14 activities with the characteristics of Tabela 44 .

The flow of activities and the calculation of the start and end week of the activity are shown in Figura 50 . The start of task A is 1, i.e. the 1st week, and it ends in week 2 (End of A = Start of A + Duration of A - 1). The start of B is equal to (1 + End of A); the end of B is (Start of B + Duration of B -1); and so on. Branching occurs when there is more than one successor task, opening up the flow into 2 or even more branches. This can be done manually, or using software such as SAS, with PROC CPM (see program at the end of this section).

Tabela 44 - Duration and successors of the activities to be programmed by the CPM

Activity	Duration (weeks)	Successor1	Successor2
A	2	B	B
B	4	C	C
C	6	D	E
D	5	F	G
E	4	H	I
F	7	I	I
G	5	J	J
H	6	J	J
I	8	N	N
J	9	K	L
K	5	M	M
L	4	M	M
M	6	.	.
N	2	.	.

The SAS program to carry out the CPM programming for this case is listed below:

```
DATA DATA;
INPUT ACTIVITY $ WEEKS SUCCESSOR1 $ SUCCESSOR2
$;
DATALINES;
A 2 B B
B 4 C C
C 6 D E
D 5 F G
E 4 H I
F 7 I I
G 5 J J
H 6 J J
I 8 N N
J 9 K L
K 5 M M
L 4 M M
M 6 . .
N 2 . .
```

```
;
PROC CPM DATA=DATA
      OUT=RESULT
      INTERVAL=WEEK
      DATE='01JAN07'D;
 ACTIVITY ACTIVITY;
 DURATION WEEKS;
 SUCCESSOR SUCCESSOR1 SUCCESSOR2;
PROC PRINT DATA=RESULT NOOBS;
RUN;
```

The result of the processing is Tabela 45 .

Tabela 45 - CPM programming results for 14 activities with SAS.

Activity	Successor1	Successor2	Weeks	E_START	E_FINISH	L_START	L_FINISH	T_FLOAT	F_FLOAT
A	B	B	2	01/Jan/07	14/Jan/07	01/Jan/07	14/Jan/07	0	0
B	C	C	4	15/Jan/07	11FEB07	15/Jan/07	11FEB07	0	0
C	D	E	6	12FEB07	25/Mar/07	12FEB07	25/Mar/07	0	0
D	F	F	5	26/Mar/07	29APR07	21MAY07	24/jun/07	8	0
E	H	I	4	26/mar/07	22APR07	26/Mar/07	22APR07	0	0
F	I	I	7	30APR07	17/jun/07	25/jun/07	12AUG07	8	0
G	J	J	5	01/Jan/07	04FEB07	30APR07	03/jun/07	17	17
H	J	J	6	23APR07	03/jun/07	23APR07	03/jun/07	0	0
I	N	N	8	18/jun/07	12AUG07	13AUG07	07OCT07	8	0
J	K	L	9	04/jun/07	05AUG07	04/jun/07	05AUG07	0	0
K	M	M	5	06AUG07	09SEP07	06AUG07	09SEP07	0	0
L	M	M	4	06AUG07	02SEP07	13AUG07	09SEP07	1	1
M			6	10SEP07	21OCT07	10SEP07	21OCT07	0	0
N			2	13AUG07	26AUG07	08OCT07	21OCT07	8	8

Where: Activity = activity name; Successor1 = first successor activity; Successor2 = second successor activity; Weeks = activity execution time in weeks; E_START = first date to start the activity; E_Finish = first date to end the activity; L_Start = last date to start the activity; L_Finish = last date to end the activity; T_Float = time off to start the activity; F_Float = time off to end the activity. Note: all times and periods are in weeks.

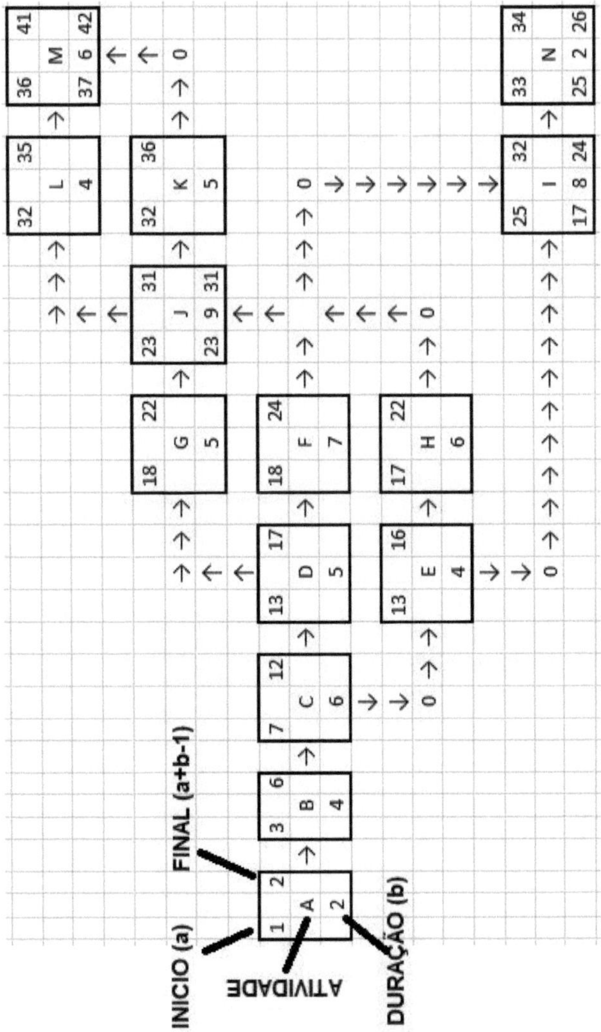

Figura 50 - Development of the schedule by the CPM for the 14 activities.

From the flowchart, a Gantt chart can be developed to illustrate the progress of activities (Figura 51).

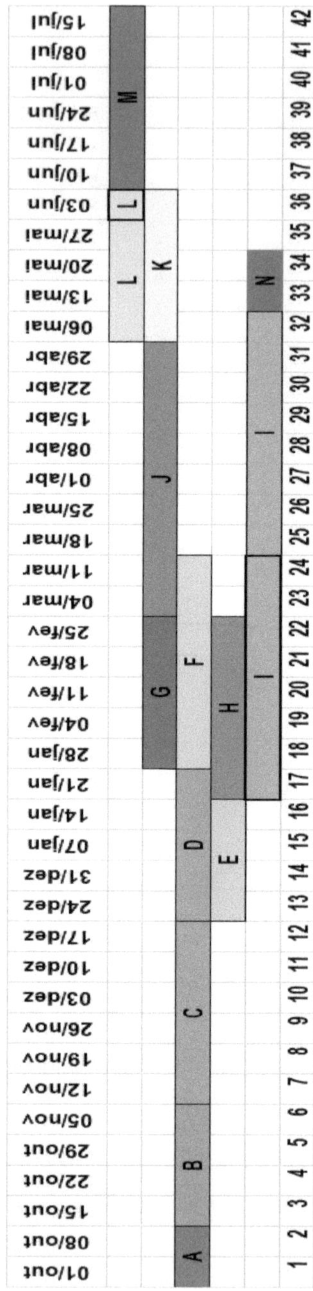

Figura 51 - Gantt chart showing the progress of the 14 activities.

The Gantt chart makes it easier to determine which activities are scheduled to start or finish (Tabela 46).

Tabela 46 - Activities with gaps, as shown by the Gantt chart.

Activity	Time	Period	Time off for
I	8	17 a 24	Start
L	1	36	Finish

13 IMPACT ASSESSMENT

The impacts of undertakings in any area of the economy are assessed in terms of their economic, environmental and social aspects.

Since the 1970s, various EIA methodologies have been developed, based on principles, criteria and indicators for different forms of environmental impact, some specific to a particular case and others more generic.

The choice of EIA method depends on several factors (CARVALHO and LIMA, 2010; IDEIA, 2002), including:

- Data and information available;
- The type of project that will be evaluated;
- Existing technical and financial resources;
- The quantity and quality of data and information available;
- The amount of time available to carry out the study;
- The Terms of Reference and the legal requirements that the EIA/RIMA must meet.
- The activities, future products, services or main processes being studied.

An attempt was made to survey and classify the most common existing methodologies for Environmental Aspect Assessment (EIA), on which the methodology proposed for application to Brazilian production of *Eucalyptus* wood for pulp and paper manufacturing was based, according to their type and purpose: for the identification and summarization phase, or for the assessment phase.

Impact assessments and risk management (next chapter) are very similar and methodologies used for one can sometimes be adapted for the other.

13.1 EIA methods for the identification and summarization phase

Its aim is to identify, qualify, quantify and present the probable impacts resulting from a project. Based on RODRIGUES (2002) and SOARES (2002), the 6 main methods were identified:

- Spontaneous Method (*Ad-Hoc*) - used for specific projects, uses *brainstorming* with the aid of tables and matrices; has no standard;
- *Checklist* method - uses standardized lists of environmental factors associated with the project; with information on forecasting techniques, impact description, value scales and weighting indices;
- Matrix methods - relate project actions to environmental factors with intersecting rows and columns; the impact of each action on each environmental factor is represented; in the Leopold Matrix, for example, with 100 project actions on the horizontal axis and 88 environmental characteristics and conditions on the vertical axis, the impacts are identified as positive or negative and quantified in terms of magnitude and importance on a scale of 0-10;
- Interaction Networks - presented in the form of diagrams, graphs or flowcharts that seek to establish the sequence of environmental impacts from a given cause and effect intervention, using graphical methods; possible impacts are associated with value parameters in terms of

magnitude, importance and probability, identifying direct and indirect impacts arising in each case;

- *Overlay* - is used for studies involving location alternatives and spatial dimension issues; it consists of over*laying* thematic maps, one for each environmental factor, to identify the environmental situation in a spatialized way;
- Simulation Model - consists of mathematical models designed to represent the structure and functioning of systems through complex relationships between quantitative or qualitative components (Physical, Biological or Socioeconomic), based on a set of hypotheses or assumptions;

13.2 EIA methods for the assessment phase

They are used to compare alternatives and can be classified according to how opinion is incorporated into the process. Based on RODRIGUES (2002) and SOARES (2002), the 3 main methods were identified:

- Quantitative methodologies - seek to associate values with the qualitative considerations formulated in the impact assessment of a project; the environment is divided into categories (such as ecology, pollution, aesthetics and human interest) divided into components and subdivided into parameters represented by a set of measures; the effects on all the parameters are expressed using the same scale, generally from 0 to 10;
- Balance Sheet - uses monetary quantification and non-quantifiable impacts are subject to qualitative analysis;
- Objective Achievement Matrix - impacts are assessed according to a cost-benefit ratio based on the weighting of the different objectives of society and the groups affected.

13.3 Principles and criteria for environmental assessment

Criteria are limits established when making judgments regarding quality or quantity, based on previously identified scientific determinations that can be controlled and which make it possible to say whether something is good or bad, beneficial or harmful, high or low cost, of great or little importance, etc. A criterion can also be understood as a way of judging compliance with a principle.

So, in order to establish criteria, it is first necessary to establish the principles, then how to judge them based on observations and study of existing data or experimentation. The criteria for verifying compliance with the principles in absolute terms (yes/no), or in relative terms (degree or percentage of compliance) must be determined by previous scientifically-based studies, or in accordance with the relevant legislation and standards.

Indicators are the (quantitative) parameters and (qualitative) attributes chosen on the basis of principles to carry out environmental analyses based on the established criteria. The ISO/TR 14061 standard (1998), translated and adapted for Brazil by ABNT as NBR 14789 (Dec. 2001), establishes the principles, criteria and indicators for the management of forest plantations, defining them as follows:

> ***Principle*** *- Legislation or fundamental rule that serves as a basis for action and is expressed in the form of an objective or attitude in relation to the function of the forest ecosystem and the relevant aspects of the social system to which this ecosystem relates.*

> ***Criterion*** *- An expression of the attitude towards parameters or requirements that reflect adherence to a principle and that relates to the state or dynamics of a system.*
>
> ***Indicator*** *- A quantitative or qualitative parameter that makes it possible to evaluate, in an objective and unambiguous way, the characteristics of the forest ecosystem or related social system, or describes elements of forest management and the production processes carried out in this system.*

According to SOARES (2002), environmental indices are mathematical functions based on two or more variables, they are the numerical expression of indicators, which in turn can be understood as:

> Tools for monitoring action strategies on the environment through systematic analysis of temporal and/or spatial deviations from a reference situation.

SOARES (2002) classifies indicators into three categories:

- **Pressure indicators** - used to assess the pressure of anthropogenic activities on the environment (example: growth in demand for pulpwood);
- **State indicators** - used to describe environmental situations (example: nitrate concentration in river water);
- **Response indicators** - used to assess efforts to solve an environmental problem (e.g. funding for soil remediation).

13.4 Classification of environmental impacts

CREA, concerned with training professionals registered with the council in environmental legal requirements, has promoted courses on the preparation of EIA/RIMA in various states; among the suggested methodologies, the course given in Rio Grande do Sul by IDEIA (2002) suggests that the impact assessment be carried out based on the characteristics of:

- IMPORTANCE - represented by the merit given to the fact in terms of its environmental effect;
- MAGNITUDE - represented by intensity in the event of an occurrence;
- SEVERITY - given by the intersection of importance and magnitude;
- FREQUENCY/PROBABILITY - represented by past occurrence or probability of occurrence;
- SIGNIFICANCE - which can be a legal requirement, a voluntary policy, or a stakeholder requirement.

The same evaluation system is suggested by the Henrique Luis Roessler State Foundation for Environmental Protection - FEPAM (2002), in Rio Grande do Sul.

The GUT method (Gravity/Urgency/Tendency) has been widely used in problem management and is described by CARVALHO (1997) as an assessment of the severity of problems (importance), the urgency of solving them (temporality) and their tendency to remain in the same condition, get worse or improve.

The "severity" characteristic of the GUT method is similar to the "severity" of IDEIA's EIA/RIMA Course.

The "urgency" of the GUT, characterized by the urgency of time, is related to the Precautionary Principle, which is the subject of the analysis for forest management certification by the FSC and of Chapter 35 of Agenda 21, "Science for Sustainable Development", which reads as follows:

> *"The precautionary approach can provide a basis for policies concerning complex systems that are not yet fully understood and whose consequences of disturbances cannot yet be predicted."*

In this case, the imminence that something serious may happen, which may have already occurred, may be happening, or may occur, determines the degree of urgency based on the Precautionary Principle, in which action must be taken in the very short term (immediately, now), in the short term (soon, in the next few days), in the medium term (soon, in the next few weeks), in the long term (in the next few months), or in the very long term (in the next twelve months); so that, if it happens, the consequences are annulled or minimized.

Precaution is also the object of analysis of liability for damage to the environment provided for in Federal Decree No. 3179/99, in its Article 41, Paragraph One:

> *"VI - failing to adopt, when so required by the competent authority, precautionary measures in the event of a risk of serious or irreversible environmental damage."*

Potential risks should therefore be characterized by their imminence as an evaluation criterion, whether due to legal, socio-economic or technical-administrative requirements.

The "tendency" characteristic of the GUT is similar to the "probability" of the EIA/RIMA of the IDEIA course, since both deal with the chances of something happening, although "tendency" also has to do with the inclination to get better or worse, and is a more complete criterion than simple probability.

The "significance" used by IDEIA is related to the "severity" of the GUT method, as it refers to the legal importance, either for the organization, or for its customers, and can be analyzed as a sub-criterion of importance, to which it is considered to be included.

In defining the environmental objectives and targets of the EMS (ISO 14000) at Alston Brasil Ltda, CHIARADIA (2001) reports that legal requirements, environmental aspects, technological, financial and operational requirements and the views of interested parties are taken into consideration; for the EIA, an environmental impact assessment matrix is used based on the Environmental Failure Mode and Effects Analysis (EFMEA) method, a variation of the Failure Mode and Effects Analysis (FMEA) method, which according to MOHR (2002) is a methodology for analyzing and discovering all the potential failure modes in a system, the effects that the failures have on the system and how to correct and mitigate the failures or effects on the system. According to the aforementioned authors, the method includes the following assessment items: the location or area where the activity is carried out, the activities, the products and/or services resulting from the activity, the expected environmental aspects and impacts, the situation of occurrence (normal or not), the incidence (frequency), class (intensity), temporality, legislation, severity (seriousness), probability and frequency, failure mode, cause of failure, and the result (positive or negative effect).

According to ROSA (1993), the assessment of environmental impacts must take into account the following factors: adversity and significance; area of occurrence (affected region, society, affected interests, etc.); intensity, severity of the impact, its reversibility and its mitigation potential; temporality (temporary / permanent), spatialization (local / regional), reversibility (reversible - returns to original conditions with the application of a mitigating or control measure / irreversible); control or mitigability.

The IDEIA course criteria, when crossed with the GUT method, as shown in Table 6, result in five analysis characteristics that have been renamed to form the so-called "**5i**" method.

The study of environmental aspects, according to Brazilian legislation, includes action planning, as do the ISO 14000 series standards (ABNT, 1996), and the ISO 9000 series standards recommend using the PDCA (Plan, Do, Control, Act) system for planning any type of management action and describe it in detail.

Tabela 47 - **Characteristics for assessing environmental aspects**

CRITERIA			RESULTING CRITERIA	
IDEA a		**GUT b**	**Original name**	**Redenomination (5i)**
Severity	Importance	Gravity	Importance	Importance
	Magnitude		Magnitude	Intensity
Frequency		-	Frequency	Incidence
Probability		Trends	Probability or Trend	Tilt
-		Urgency	Urgency	Imminence
Significance		Gravity	Importance	-

Sources: (a) IDEIA (2002); (b) CARVALHO (1997).

Action planning involves making decisions about the actions to be taken. In this respect, the Decision Process Methodology, as described by CARVALHO (1997), helps.

According to the above, the "5i", Decision Process and PDCA methodologies are complementary, as the first helps to assess the degree of impact, the second to decide on the measures to be taken and the third to plan actions, which are the general steps to be followed in the study of environmental aspects in accordance with the ISO 14000 series standards and in accordance with Brazilian legislation on EIA/RIMA. However, the scope of this work is restricted to the assessment of environmental impacts, with the classification and methodology set out below.

13.4.1 Qualitative classification

As for the Genre: natural and anthropic. The natural aspect is divided into biotic and physical (abiotic). The anthropic aspect has four subtypes, according to the ISO

14000 series (ABNT, 1996a/b), known as the classification as to significance, which is the classification as to who is significant the effect resulting from the impact considered, which are: 1) legal requirement; 2) voluntary policy; 3) stakeholder requirement or 4) requirement due to pressure from society. Although this type of classification is relevant, the subtypes mentioned can be considered part of the cultural anthropic class. In this study, the classification in terms of genre will be restricted to the types: natural (biotic and physical) and anthropic (social, cultural and economic).

- **Mode**: direct and indirect (SOARES, 2002a).
- **As for the result**: positive and negative (SOARES, 2002a).
- **Reversibility**: reversible, partially reversible, irreversible.
- **As for the result**: emergency or unforeseen; wear and tear or maintenance failure; system or operational failure; equipment or design failure; resulting from the activity itself.

Impacts can also be classified **according to the type of cause**, which can be: matter, energy, living beings, processes or phenomena, but this analysis will not be carried out in this study, as it can be considered implicit in the activity that generates the effect, and is only complementary information.

As for Mitigability: mitigable - the cause and/or negative effect can be minimized or eliminated without having to permanently stop carrying out the productive activity; non-mitigable - in order for the adverse cause or effect to be minimized or eliminated, it is necessary to permanently stop carrying out the productive activity.

Detectability - According to SOARES (2002-b), potential and actual environmental impacts should be assessed in terms of how easy they are to detect (detectability) using the monitoring means available, according to the following criteria: difficult (unlikely that the environmental impact will be detected), moderate (likely that the environmental aspect will be detected within a reasonable period of time) and easy (practically certain that the environmental impact will be detected quickly).

13.4.2 Absolute ranking

Spatial - this is done by the area of coverage: local, regional, global; it can be given in units of surface measurement (m^2, ha, km^2) or, when the measurements of the affected part and the whole considered are known, it can be given as a proportion of the whole under study. According to SOARES (2002a), impacts should be considered in the following spatial ranges: restricted to the place of occurrence; restricted to the physical boundaries of the company; affecting the region adjacent to the company; affecting large areas outside the company.

Temporal (or grace period of the effect) - this is done by the period in which the effect remains active or present, which can be temporary (short, medium or long term) or permanent; it is given in units of time (hours, days, weeks, months, years). SOARES (2002b) adds that the temporality aspect must take into account whether the impact is the result of activities, products and/or services developed: in the past which still generate environmental impacts (environmental liabilities); in the present; or which are being implemented or may have an impact in the future.

13.4.3 Relative ranking

The parameters for quantifying environmental aspects are graded on a scale according to their degree of impact, for example: graded from 0 to 10. In this case, the minimum and maximum degrees must be determined for each factor considered and the intermediate values must be calculated according to parameterization methods.

Gravity or Importance - This is the relative value of an event, or severity if it occurs, in relation to the overall value of the whole considered. The greater the importance, the greater the value of the event over the value of the whole to which it relates. It is the risk of damage estimated on the basis of what has already occurred. Importance has to do with the size of something in relation to the average: minimal, small, medium, large, exceptional. An example of severity classes found in the literature is given in Table 7.

Urgency or Imminence - This is the time remaining for action to be taken on an event that may occur. It is determined by the imminence of an event occurring immediately, in the medium term, or in the remote term. Imminence refers to the temporal situation of an event itself. It represents the urgency of taking action in relation to something that could happen. It is the degree of urgency to adopt a measure. An example of the degree of urgency, drawn up in 1992 for deadlines to combat ants in Bahia Sul, is shown in Table 8 below.

Tabela 48 - Example of severity categories

Definition	Category	Description
Catastrophic	I	Death, loss of system or severe environmental damage.
Review	II	Serious injury, serious occupational illness, major damage to the system or the environment - Significant consumption of natural resources; high generation of pollution.
Marginal	III	Minor injuries, non-important occupational illnesses, minor damage to systems or the environment - Moderate consumption of natural resources; moderate generation of pollution and waste
Despicable	IV	Less than minor injuries, non-important occupational illnesses or no damage to systems or the environment - Negligible consumption of natural resources; no significant pollution.

Source: SOARES (2001).

Tabela 49 - Example of time levels - Deadline and actions to take

CONTENTS	RISK	ACTION	DEADLINE
0 - 10	None	Monitoring	12 months
11 - 25	Very small	Monitoring	6 months
26 - 45	Small	Monitoring	3 months
46 - 75	Medium	Combat	1 month
76 - 100	High	Combat	15 days
>100	Very high	Combat	Immediate

Source: FLORIANO (1998c).

Intensity or force involved - This is the degree to which a given fact occurs in relation to its pattern. When it is parametric and its distribution is normal, it can be classified in terms of mean and standard deviation. When it is a non-parametric attribute, it can be quantified in percentages and transformed into a continuous variable, if the occurrence is greater than 30 times, by appropriate mathematical methods, so that it can be treated as a parameter. According to SOARES (2002-b), environmental impacts should be assessed according to their criticality in relation to the environment as severe (adverse environmental impact causing irreversible, critical or difficult-to-reverse damage and/or endangering the lives of human beings outside the company, slight (adverse impact causing reversible or circumventable damage and/or threatening the health of human beings outside the company) and minimal damage (when the environmental impact causes minimal or imperceptible damage).

Frequency or Incidence - This is the number of times an event occurs per unit of time; it can also be a proportion in relation to its normal pattern of occurrence. An example of frequency classes is shown below:

Tabela 50 - Example of frequency levels

Definition	Level	Description
Frequent	A	Occurs frequently (or high probability), or occurs permanently when the activity starts.
Likely	B	It will occur several times in the life of the system or item.
Occasional	C	It will occur a few times over the life of the system or item.
Remote	D	It is not expected to occur (although there is some expectation) over the life of the item or system.
Unlikely	E	It can be assumed that it will not occur throughout the life of the system or item.

Source: SOARES (2001).

Tendency or inclination - This is the prognosis that something will remain the same, improve or worsen over time, without anything being done about it. It relates to the risk of future damage. The greater the risk of the situation getting worse, the higher the grade that should be assigned; the greater the tendency for the situation to improve, the lower the grade. Consider a working machine: over time there is a tendency for it to wear out and for its condition to deteriorate; if it is a good quality machine and there is permanent and adequate maintenance, the tendency for its operating conditions to worsen is very slow and it can be considered stable for a relatively long period of time. In another situation, a farm is being analyzed and physically planned: the trend is for improvement with the appropriate use of the spaces; areas used for agriculture that present problems and whose future use planning is to restore the original forest, have a tendency to improve to the maximum degree, while flat and fertile areas used with cultivated forests have a slight tendency to worsen, even if appropriate management is adopted, as agricultural crops are more damaging to the environment than forests; by taking the weighted average of the trends of each area in relation to its future destination, the average trend for the property as a whole can be inferred.

13.4.4 Sizing up impacts

According to SOARES (2002-b), the final significance of environmental impacts and the actions to be taken according to the degree of impact should be measured according to Table 10.

Tabela 51 - Final significance of environmental impacts and actions to be taken

Compliance with legislation, stakeholders and global concerns	Score obtained	Significance	Action to be taken
	0 to 6 points	Despicable	"Maintain routine" (if the respective environmental aspect is real) or "Emergency plan" (if the respective environmental aspect is potential)
- If there is applicable legislation or demand from interested parties, the significance should be taken as at least "substantial".	8 to 16 points	Meaningful	"Operational control" (if the respective environmental aspect is real) or

Compliance with legislation, stakeholders and global concerns	Score obtained	Significance	Action to be taken
- If there is applicable legislation and it is not being complied with, the significance should be taken as "important".			"Emergency plan" (if the respective environmental aspect is potential)
- If none of the above situations occur, consider the score obtained to define significance.	18 points or more	Important	"Operational control" (if the respective environmental aspect is real) or "Emergency plan" (if the respective environmental aspect is potential)

Source: SOARES (2002-b).

The classification of risks in the GUT method (CARVALHO, 1997) is made in accordance with Tabela 52 .

Tabela 52 - Degree of comfort with decisions

Degree of comfort		Probability	Gravity
I have no comfort	↓	High	High
I have no comfort	↓	High	Average
I'm comfortable	↑	High	Low
Watch out	↔	Average	High
Watch out	↔	Average	Average
I'm comfortable	↑	Average	Low
I'm comfortable	↑	Low	High
I'm comfortable	↑	Low	Average
I'm comfortable	↑	Low	Low

Probability: High - will happen, Medium - may happen, Low - unlikely to happen; Severity: High - very important, Medium - important; Low - negligible. Source: CARCALHO (1997).

By crossing the indications in the tables of significance of impacts and comfort in decisions and adjusting the results obtained by crossing, the following can be obtained Tabela 53 for classifying the relevance of environmental impacts.

Tabela 53 - Classification of the degree of relevance or impact using the "5i" method .

Value Classes	Relevance or Impact	Observations
0 a -2,0	LOW	There is comfort in carrying out the project without major risks. Some mitigation measures may be required.
2,1 a -4,0	LOW	It calls for caution. It is necessary to adopt some mitigating measures for the activities with the greatest impact.
4,1 a -6,0	MEDIUM	Can be carried out under strict control. Requires mitigation and compensation measures for most activities.
6,1 a -8,0	HIGH	It is a high-risk project. It requires extreme care in its execution and major actions to reduce risks.
8,1 a -10,0	HIGH	It is only justified in cases of extreme social need. The risks are too high. It tends to be unfeasible.

Source: Adaptation, with the aid of the classes for determining the significance of environmental impacts suggested by SOARES (2002-b) crossed with the classes for assessing potential risks of the GUT method described by CARVALHO (1997), to quantify the degree of total impact, or relevance.

14 RISK MANAGEMENT

14.1 Introduction

Organizations face internal and external factors and influences that make it uncertain whether and when they will achieve their objectives; the effect of this uncertainty on an organization's objectives is referred to as "risk" (ISO, 2009).

Risk can be conceptualized as the possibility of an event occurring that is adverse to the achievement of an objective. It is the effect of uncertainty on objectives.

According to ISO (2009), organizations must create a risk management framework, implement it and integrate it into their management and culture, maintain a risk monitoring system and continually improve plans to contain the risks to which they are subject. Managing risks in an organization has the main effect of improving operational effectiveness and efficiency and health and safety performance, as well as environmental protection. The risk management framework is a set of components that provides the basis and organizational arrangements for the design, implementation, monitoring, review and continuous improvement of risk management throughout the organization.

According to the ISO 31.000 standard, the risk management process should be carried out according to the Figura 52 . In establishing the risk context, an organization must articulate its objectives and define the external and internal parameters to be taken into account when managing risks, defining the scope and risk criteria for the remaining process.

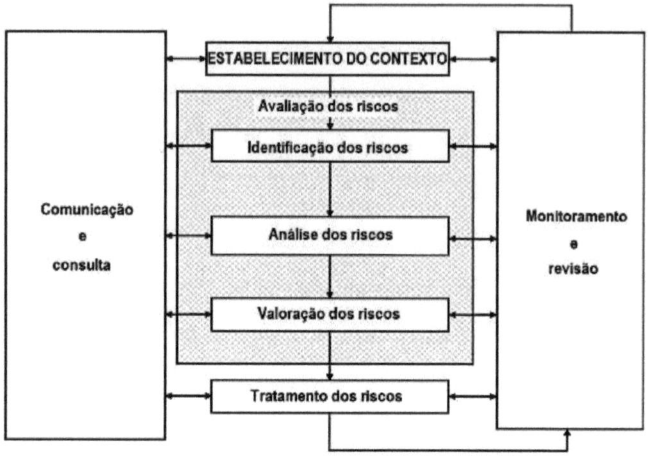

Figura 52 - Risk management process (ISO, 2009).

Risk treatment can involve (ISO, 2009):

- Avoiding the risk by deciding not to start or continue with the activity that gives rise to the risk;
- Taking or increasing risk in order to pursue an opportunity;
- Remove the source of risk;
- Change the probability of it happening;
- Change the consequences;
- Sharing risk with stakeholders (including contracts and risk financing);
- Retain risk by informed choice.

Forest production planning has traditionally been based on the assumption that the outcome of the management activity is certain. In reality, the results of forest management are uncertain and subject to potential risks caused by the environment, man and the trees themselves, due to the long periods needed for the trees to mature and be ready to be harvested. The management of forest areas must take these risks into account. Risk analysis should be part of all forestry decisions, bearing in mind that it is almost impossible to obtain a complete understanding of all the alternatives (BETTINGER, 2009).

In the past, it was customary to recommend that a loss of 10% of production be considered. Therefore, 10% was added to the area needed for industrial supply in order to avoid the risk of a shortage of wood due to possible losses. Since the development of forest management with improved techniques, this has ceased to be a viable practice, as it is possible to manage risks in such a way as to reduce the planting area to what is strictly necessary, or to accurately determine the losses that may occur, taking them into account when planning production.

14.2 Risk classification

Oliveira (2009) classifies risks as strategic (Tabela 54), compliance (certification) (Tabela 55) and operational (Tabela 56).

Tabela 54 - Examples of strategic risks (adapted from Oliveira, 2009)

DANGERS to business*	Effects	Probability of happening (1-5)
Climate change	Change in productivity; Changes in the cost of factors of production; Increased volatility;	Variable
New markets	Changes in demand and supply of raw materials and products;	Variable
Social framework	Social upheaval / strikes;	Variable
Public policies	Changes in legislation.	Variable

Source: Oliveira (2009). Where: (*) Hazards = value, quality and quantity of product or products.

Tabela 55 - Examples of Compliance Risk

Risk Class	Related problems	Probability of happening (1-5)
Administrative regulation	Delayed project deadline	Variable
	Increased administrative costs	
	Increased financial costs	
Certification	Non-certification of the final product	Variable
	Certified wood stock-outs	
	Changing prices and market supply conditions	

Source: Oliveira (2009).

Tabela 56 - Examples of operational risks

Phase/Activit y	Risks related to operations	Impact on value (1-5)	Critical factors
Installation	Plant adaptation	2 a 4	Plant adaptation and nutritional recommendations
	Wrong fertilization	1 a 5	
	Poor preparation	3 e 4	

Phase/Activity	Risks related to operations	Impact on value (1-5)	Critical factors
	Plant quality and number	2 e 4	Number of plants and quality of planting and genetic diversity
	Quality of planting	2 a 5	
Maintenance	Interspecific competition	1 a 4	Health
	Wrong fertilization	1 a 4	
	Biotic agents	2 a 4	Managing interspecific and intraspecific competition
	Time of cut	1 a 2	
	Number of courts	2 a 4	
Exploration	Quality of the material to be explored	1 a 3	Information management
	Quality of cut and filling	1	
	Land and road density	1	Legal compliance and best practices
	Non-conformities	2 a 4	
Erosion	Destruction of infrastructure	4	Road network management and post-fire operation
	Soil loss after a fire	2 a 5	
Livestock, theft or vandalism	Loss of wood	1 a 2	Anticipating local tensions
	Destruction of plantations,	4	Selection of PS and F
	Theft of factors of production	3	
Abiotic events	Forest fire	1 a 5	Managing the time of cutting/harvesting
	Drought	1 a 4	
	Frost Other	2 a 4	Understory management and combat tactics

Source: Oliveira (2009).

14.3 Identifying and quantifying risks

The main risks in forestry production are listed below:

- Environmental
  - Environmental disasters (wind, lightning, drought, flooding).
  - Fire.
  - Pests.
  - Diseases.
  - Soils - compaction, depletion, contamination, erosion.
  - Biological contamination.
- Social
  - Political - social political instability.
  - Legal - changes in legislation (loss of productive area).
  - Social instability - environmental pressures, strikes;
- Economic
  - Interest rate - increase.
  - Economic instability - internal and external.
  - Taxes - increase.
  - Surplus industrial stock.
  - Price of products.
  - Input costs.
  - Production costs.
  - Uncontrolled inflation.
  - Floating exchange rate.
  - Indebtedness.
  - Reducing consumption.
  - Replacing products with similar ones.

In risk analysis, two or more factors can be adopted to parameterize them in order to carry out the quantitative analysis. Oliveira (2009) suggests quantification according to probability (at levels 1=rare to 5=very likely) and impact if it occurs (at levels 1=not important to 5=very important). Parameterization can also be carried out on the basis of, for example, the GUT method (severity, urgency and trend), or the 5i method (importance, imminence, incidence, intensity, inclination), where 3 of the 5i factors coincide with the 3 of the GUT, with the 5i being more comprehensive. The 5i method was developed to classify the risks of environmental impacts in forestry projects. The parameterization of risks using the method is shown in Tabela 57 a Tabela 61 . For quantification, it is not necessary to use all the factors (the 5i), but at least the 3 that coincide with the GUT. The index values for each factor selected must be added together; this sum is divided by the number of factors used. The index values for each factor range from 1 to 10 and the average will therefore also be a value from 1 to 10, increasing the risk the closer the parameterization result is to 10. The factors in the 5i method are described below:

1) Importance or seriousness - This is the relative value of a fact, should it occur, in relation to the overall value of the whole considered. It must be estimated on the basis of facts that have occurred.
2) Imminence or urgency - This is the time remaining before action is taken in relation to an event that may occur. It represents the urgency of time to take action on something that could happen.
3) Intensity - This is the potential damage that a normal average quantity can cause to living beings, or the strength with which a certain event occurs in relation

to its average pattern.

4) Incidence - The frequency with which something occurs on a scale that must be transformed into 1 to 10.

5) Slope or trend - This is the prognosis of whether something will remain the same, improve or worsen over time, considering the current conditions and state.

Tabela 57 - Importance Classes

Contents	Damage/ All (%)	Criteria
1	≤10%	Insignificant damage or risk to human health and other forms of life or ecosystems. Negligible damage or risk to production systems. Changes to up to 10% of the landscape. Production losses of up to 10% for a period not exceeding 30 days.
2	>10 e ≤ 20	Risk of very minor injuries. Alteration of landscape or habitats by up to 20%. Very minor damage or risk to productive activities. Loss of production of up to 20% for a period not exceeding 30 days.
3	>20 e ≤ 30	Risk of minor injury. Alteration of landscape or habitats by up to 30%. Minor damage or risk to productive activities. Loss of production of up to 30% for a period not exceeding 30 days.
4	>30 e ≤ 40	Risk of minor injury. Risk of chronic illness while carrying out the activity. Alteration of the landscape or habitats by up to 40%. Minor topographical changes. Minor damage or risk to productive activities. Production losses of up to 40% for a period not exceeding 30 days.
5	>40 e ≤ 50	Risk of medium injuries. Risk of chronic illness with minor permanent sequelae. Alteration of landscape or habitats by up to 50%. Topographical changes with minor consequences for soils and water regime. Medium damage or risk to productive activities. Production losses of up to 50% for a period not exceeding 30 days.
6	>50 e ≤ 60	Risk of medium injuries. Risk of chronic illness with permanent sequelae of medium severity. Alteration of the landscape or habitats by up to 60%. Topographical changes with medium

Contents	Damage/ All (%)	Criteria
		consequences for the soil and water regime, requiring mitigation measures at a cost of up to 5% of the project. Risk of interrupting productive activities. Production losses of up to 60% for a period not exceeding 30 days.
7	>60 e ≤ 70	Reduction of habitats by up to 70%. Risk of accidents and serious diseases. Topographical changes with consequences for the soil and water regime, requiring mitigation measures at a cost of up to 10% of the project. Risk of considerable loss of production systems. Risk of interruption, for a long period, of any process. Production losses of up to 70% for a period not exceeding 30 days.
8	>70 e ≤ 80	Risk of putting one or more species in danger of local extinction. Reduction of habitats by up to 80%. Risk of accidents and very serious diseases. Topographical changes with consequences for the soil and water regime, requiring mitigation measures at a cost of up to 15% of the project. Risk of almost complete loss of the production system. Risk of interruption, for a very long period, of any process. Production losses of up to 80% for a period not exceeding 30 days.
9	>80 e ≤ 90	Risk of placing one or more species in danger of local extinction. Reduction of habitats by more than 80%. Risk of accidents and very serious diseases. Topographical changes with consequences for the soil and water regime, requiring mitigation measures at a cost of up to 20% of the project. Risk of almost complete loss of the production system. Risk of interruption, for a very long period, of any process. Production losses of up to 90% for a period not exceeding 30 days.

Contents	Damage/ All (%)	Criteria
10	>90%	Risk of extinction of one or more species or habitats on site. Risk of death. Topographical changes with consequences for the soil and water regime, requiring mitigation measures costing more than 20% of the project. Risk of complete loss of the production system. Risk of permanent interruption of any process. Production losses of more than 90% for a period of no more than 30 days.

Source: Adaptation of the importance index tables found in the bibliography cited.

Tabela 58 - Imminence classes

Contents	Deadline for action (days)	Criteria
1	720	Remote risk of minor problems after 2 years.
2	360	Small risk of minor problems after 1 year.
3	180	Medium risk of minor problems after 6 months.
4	90	Medium risk of medium problems after 3 months.
5	60	Medium risk of medium problems after 2 months.
6	45	High risk of medium-severity problems after 45 days.
7	30	High risk of serious problems after 30 days.
8	15	High risk of serious problems after 15 days.
9	5	High risk of very serious problems after 5 days.
10	1	Very high risk of very serious problems at any time.

Source: Adaptation of the imminence index tables found in the bibliography cited.

Tabela 59 - Intensity classes

Contents	Criteria				
	Domestic electricity (amperage)**	**Human physical losses* (%)**	**Remaining specimens by species (thousands)**	**Risk of annual deaths (nº/100mil)**	**LD50 of toxicants (mg/Kg)**
1	≤ 0.5mA	≤ 1	≥ 200	≤ 7	≥ 10000
2	> 0.5 to ≤ 1mA	>1 e ≤ 3	>200 e ≤ 150	>7 e ≤ 11	>8000 e ≤ 10000
3	> 1 to ≤ 2mA	>3 e ≤ 6	>150 e ≤ 80	>11 e ≤ 17	>6000 e ≤ 8000
4	> 2 a ≤ 5mA	>6 e ≤ 10	>80 e ≤ 60	>17 e ≤ 20	>4000 e ≤ 6000
5	> 5 to ≤ 10mA	>10 e ≤ 15	>60 e ≤ 40	>20 e ≤ 23	>2000 e ≤ 4000
6	> 10 to ≤ 15mA	>15 e ≤ 20	>40 e ≤ 20	>23 e ≤ 27	>1000 e ≤ 2000
7	> 15 to ≤ 20mA	>20 e ≤ 30	>20 e ≤ 10	>27 e ≤ 30	>500 e ≤ 1000
8	> 20 to ≤ 30mA	>30 e ≤ 40	>10 e ≤ 5	>30 e ≤ 35	>50 e ≤ 500
9	> 30 to ≤ 50mA	>40 e ≤ 50	>5 e ≤ 2	>35 e ≤ 40	>5 e ≤ 50
10	> 50mA	> 50	< 2	> 40	< 5

(*) Physical losses of the human being, in percentage: reduction of senses - sight, hearing, touch, smell and taste; reduction of physiological functions; reduction of vital organs or their potential; loss or paralysis of limbs with reduction of productive capacity; etc. (**) Adaptation for exposure of 10 seconds according to the standard: IEC/TR2 60479-1 (IEC, 1994). Source: Adaptation of the intensity index tables found in the bibliography cited.

Tabela 60 - Incidence Classes

Contents	Criteria			
	Normality*	Frequency Annual	Freq. %	Criteria
1	≤(X-2S)	≤ 0,5	≤10%	Rare occurrence, may not happen.
2	>(X-2S) and ≤(X-1,5S)	>0,5 e ≤ 1	>10 e ≤20	Remote occurrence.
3	>(X-1,5S) and ≤(X-1S)	>1 e ≤ 3	>20 e ≤30	Scarce occurrence.
4	>(X-1S) and ≤(X-0.5S)	>3 e ≤ 5	>30 e ≤40	Occasional occurrence.
5	>(X-0.5S) and ≤(X)	>5 e ≤ 7	>40 e ≤50	Small occurrence.
6	>(X) and ≤(X+0.5S)	>7 e ≤ 10	>50 e ≤60	Medium occurrence.
7	>(X+0.5S) and ≤(X+1S)	>10 e ≤ 20	>60 e ≤70	Medium to high occurrence.
8	>(X+1S) and ≤(X+1,5S)	>20 e ≤ 40	>70 e ≤80	High occurrence.
9	>(X+1,5S) and ≤(X+2S)	>40 e ≤ 50	>80 e ≤90	Very high occurrence.
10	>(X+2S)	> 50	>90%	Very high or permanent occurrence.

(*) Normality: X=mean; S=standard deviation. Source: Adaptation of the incidence rate tables found in the bibliography cited.

Tabela 61 - Slope classes

Contents	Trends	Chances (%)	Criteria
1	to Improve	>75%	Strong tendency to improve.
2		>50 e ≤75	Reasonable tendency to improve.
3		>35 e ≤50	Moderate tendency to improve.
4		>0 e ≤35	Slight tendency to improve.
5	Stability	100%	There is no tendency for it to get worse, or better, if nothing is done.
6	of Getting Worse	>0 e ≤ 35	Slight tendency to get worse.
7		>35 e ≤ 50	Moderate tendency to worsen.
8		>50 e ≤ 75	A reasonable tendency to get worse.
9		>75 e ≤ 90	High tendency to get worse.
10		>90%	It's bound to get worse.

Source: Adaptation of the slope index tables found in the bibliography cited.

The result of the risk analysis leads to an interpretation of the parameterization result, which can be carried out as follows Tabela 62 .

Tabela 62 - Interpretation of the parameterization result using the 5i method.

Value Classes	Degree of risk	Observations
0 a -2,0	LOW	There is comfort in carrying out the project without major risks. Some mitigation measures may be required.
2,1 a -4,0	LOW	It calls for caution. It is necessary to adopt some mitigating measures for the activities that pose the greatest risk.
4,1 a -6,0	MEDIUM	Can be carried out under strict control. Requires mitigation and compensation measures for most activities.
6,1 a -8,0	HIGH	It is a high-risk project. It requires extreme care in its execution and major actions to reduce risks.
8,1 a -10,0	HIGH	It is only justified in cases of extreme necessity. The risks are too high. It tends to be unfeasible.

Based on the results of the risk analysis, actions must be defined to minimize, eliminate or contain the risks if they occur, which is the next phase of risk treatment.

14.4 Risk treatment

Risk treatment is characterized by developing actions to modify risks, eliminating, minimizing or creating contingency plans in the event of unwanted effects, as well as implementing these actions.

Risk treatment is an ongoing process in which the treatments applied are monitored, quantified to check that the level of risk is tolerable, evaluated for the effectiveness of the treatment and new risk treatment generated based on the results.

According to ISO 31.000/2009, risk treatment options are not necessarily mutually exclusive or appropriate in all circumstances, and may include (ISO, 2009):

> "a) avoid the risk by deciding not to start or continue with the activity that gives rise to the risk;
>
> b) taking or increasing risk in order to pursue an opportunity;
>
> c) remove the source of risk;
>
> d) change the probability of occurrence;
>
> e) change the consequences;
>
> f) sharing risk with stakeholders (including contracts and risk financing); and
>
> g) retain the risk by informed decision."

14.5 Financial risk analysis (Monte Carlo method)

The Monte Carlo method has been used to analyze risk in forestry projects. The method is based on simulating different production scenarios with different financial results. As the simulation can include different planning horizons, it is advisable to carry out the simulation with an infinite series indicator such as EAV (equivalent annual value) or ETV (expected land value). It is common to find studies using the NPV or the rate of return, but this is not the most appropriate as it is determined for finite series.

The simulation involves changes to all the resources and other factors that influence their value, including all the possibilities for variation in each one, such as the exchange rate, an increase in the minimum wage, loss of forestry productivity, an increase in the price of inputs and machinery and all other unforeseen events with economic repercussions on the project. Every factor with inherent uncertainty must be taken into account in the analysis. The simulations must involve scales that cover the entire range of variation of each resource. Generally, hundreds or thousands of simulations are carried out and the financial result is calculated for each one, be it VET or VAE, or another chosen financial index. This requires the use of computers and appropriate programs. Once the simulations have been carried out and the financial results obtained for each one, the percentage of unacceptable results is checked; this percentage is the risk of the venture not being financially successful.

The simulation can be carried out randomly on the variation of each factor or resource within its range of variation and its probability of occurrence, simulating the tasks randomly based on the probabilities of occurrence of each event, thus seeking to get as close as possible to the real situation.

15 MONITORING

15.1 Introduction

Controlling activities, costs, quality, etc. has been a concern ever since the first organization was created. New control tools emerge every day and organizations are always changing their systems.

Monitoring is a new word in Portuguese dictionaries and means supervising activities to ensure that they are on course and within the planned schedule and parameters to achieve objectives and performance targets.

Monitoring is closely related to PDCA (MIRSHAWKA , 1990), the acronym for Plan, Do, Check, Action, the cycle of Dr. Shewhart, known in Japan as the cycle of Dr. Deming, who spread it in the country. PDCA was modified and called MAMP (Process Analysis and Improvement Method) by SEBRAE (GRIMALDI and MANCUSO, 1994). It is a methodology for improving quality that involves the four steps: Plan, Execute, Evaluate and Correct, which are the basis for developing a monitoring system.

Everything that involves measurement and data is somehow related to monitoring. One type of monitoring used in industry is statistical process control, another is environmental control.

The standards of the ISO 9000 family deal with this issue, and it is one of the requirements for certification. According to NBR ISO 9001, the supplier must first identify which statistical techniques are necessary for establishing, controlling and verifying the capability of its processes and the characteristics of its products; then it must establish procedures for implementing and controlling their application. So far, we have covered the first three steps of the PDCA, and the last step, *Action*, is dealt with in the corrective and preventive action requirement.

Monitoring can be more than simply measuring and evaluating, as it might seem at first glance. The scope of the verb monitor, which is not found in most dictionaries from before the 1990s, affects the entire PDCA, depending on its scope.

Monitoring systems can be classified into three main types:

- Evaluation monitoring - corresponds to the PDCA *check*;
- Control monitoring - covers evaluation and correction;
- Monitoring for continuous improvement - covers planning, execution, evaluation and correction.

Monitoring for evaluation is when a certain activity is measured periodically and presented in tables and graphs for simple monitoring. The end result is usually a periodic report of averages and variances, and sometimes trends. It mainly serves to verify the current situation and make predictions based on this.

Monitoring for control is mainly represented by statistical process control, where the activity is monitored to check that it is within certain pre-established parameters. When variations outside the limits occur, actions are taken to correct them. Its main objective is to prevent processes from getting out of control by keeping them within pre-defined standards, resulting in products and services of the same quality on an ongoing basis.

Monitoring for continuous improvement is monitoring in which the results of measurements or evaluations are used for calculations that will determine actions to be taken based on the results obtained. The aim of this type of monitoring is to automate stages or entire processes, eliminating the human factor in decision-making. They are usually adopted to reduce costs, as human beings have a habit of exaggerating doses in order to be sure of the results. For example, when a worker finds a large anthill and calculates the dose of ant killer, he may apply more than the calculated dose for fear that it won't be enough.

When developing a monitoring system for continuous improvement, it is necessary to gather all the information on the subject and follow the four stages of the PDCA: planning, execution, evaluation and correction. These stages are related to the phases of the decision-making process as shown in Tabela 63 . Monitoring for continuous improvement can be considered a fusion of the two methodologies: PDCA and Decision-Making Process.

It starts with gathering information, describing and analyzing the situation itself and the problem or problems it presents, then planning, defining the objectives of the system to be developed and analyzing the possible alternatives for solving the problems in order to achieve the objectives, defining what and how to measure the variables of the problem periodically and what to do in each case, taking into account all the alternatives. Then we move on to execution, measuring and verifying the occurrences and making the decisions as planned and, finally, we make the necessary corrections to the system and to the solutions applied to the problems that have occurred and that have not had the desired effect, planning the execution of the corrective actions, which is already the start of a new cycle.

Tabela 63 - Stages of Monitoring x Stages of the Decision-Making Process

Development stages	Decision-making process			
	Situation Analysis	Problem Analysis	Decision Analysis	Analysis of Potential Problems
Information gathering	x	x		
1ª stage Planning			x	x
2ª stage Execution			x	
3ª stage Evaluation	x	x		
4ª Corrections stage			x	x

In order to develop a monitoring system for continuous improvement, it is advisable to first study the subject of the "Decision Process" and its stage called "Decision Analysis", the latter of which is briefly presented in the section below.

15.2 Decision analysis

When we have a problem and we know the causes of the problem, we may have different solutions, i.e. actions we can take to solve the problem. The first thing we need to do at this point is determine the purpose of our decision.

We often make decisions based on previous cases. This limits us to past events. We need to look to the present and future when making a decision in order to avoid the mistakes of the past. Setting the criteria for a decision correctly allows us to take a broad view. "It's always worked out that way" is not a welcome phrase in this case.

15.2.1 Purpose of the decision (Decision Objective)

The purpose of the decision must be formulated through an **action verb** that stimulates something, a **target object** referring to the action of the verb that indicates what is being stimulated and one or more **modifying agents** that indicate limitations of the decision, where the action verb, the target object and the modifying agents describe the focus of our decision.

When establishing a decision purpose, we must take into account that it must allow us to produce a set of criteria (objectives) and a group of comparable alternatives.

The components of the purpose of a decision:

- **a) action verb** - indicates how the alternatives will be approached;
- **b) target object** - identifies the general group of alternatives to be considered;

- **c) modifying agents** - limit the group of alternatives, refine the purpose of the decision and highlight the need to make a decision.
- **Focus of the decision** = a + b + c.

15.2.2 CRITERIA

The criteria we have for making a decision are our desires, the restrictions we make, the objectives we have in relation to the purpose of the decision, in other words, they are the prerequisites we have for making the decision in relation to our purpose.

The criteria we have to decide on can be **mandatory**. These we don't negotiate and are eliminatory. Others may not be and can be negotiated: these are the **desirable** criteria. We need to separate them as in Tabela 64 .

Tabela 64 - Classification of decision criteria

Mandatory	Desirable
They are indispensable	Relative importance
They are measurable	They can be measurable
They're realistic	All those that are not mandatory
The following are essential	They are negotiable

The mandatory criteria are the minimum we want. Failure to meet any of them is a factor in eliminating alternatives. The desirable criteria are not eliminatory and some are more desirable than others, so we need to assign weights to them in order to rank them.

The criteria can be understood as the objectives we want to achieve with the decision to be made.

15.2.3 Alternatives

The alternatives we need to decide on may be obvious and present in the market or the environment in which we find ourselves. But often, not even past experience exists and we have to produce them. Again, we can use *brainstorming* to generate alternatives as a group.

Even if we have several alternatives, it's important that we make use of the information we already have to produce them, or complement our group of alternatives:

Using the purpose of the decision

- One of the functions of the decision purpose is to delimit and define a group of alternatives; in the decision purpose we can find some alternatives, or an indication of how to find them;
- Break down the purpose of the decision into sub-purposes, each of which can be used to generate a sub-alternative, look for ways to combine the sub-alternatives into a single complete one.

Use of (objective) criteria

- the criteria create limitations and give an indication of the alternatives we have;
- ask, for each criterion, what ways we have of satisfying it by generating alternatives, then try to combine them to find a complete solution.

Use of adverse consequences

- Consider the adverse consequences of improving an existing alternative or suggesting new ones.

15.2.4 COMPARISON OF ALTERNATIVES

The alternatives must first be compared against the mandatory criteria and then against the desirable ones.

ALTERNATIVES TO THE MANDATORY CRITERIA

In this case, we will simply eliminate the alternatives that do not meet one of the mandatory criteria. The alternatives that pass the test will have to be classified in the following 5ª step.

A comparison grid can be used to facilitate comparison, as follows Tabela 65 .

Tabela 65 - Alternatives vs. mandatory criteria

Criteria (description)	Alternative 1		Alternative 2		etc.	
	Attend	No answer	Attend	No answer	Attend	No answer
First criterion						
Second criterion						
etc.						

ALTERNATIVES TO THE DESIRABLE CRITERIA

Alternatives should be given scores from 0 to 10 for meeting each of the desirable criteria, starting with a value of 0 (zero) for those that do not meet the criterion considered, up to a value of 10 for those that fully meet it. The sum of the products of the criteria's weights and the alternative's score will give us the number of points the alternative has achieved. The more points, the better the alternative (Tabela 66).

Two or three alternatives should be chosen from among those with the highest number of points. The choice of the number of alternatives will depend on our experience and common sense.

Tabela 66 - Alternatives x Desirable criteria

Criteria		Alternative 1		Alternative 2		ETC...	
Description	Weight (a)	Note $(b)_1$	Points $(a \cdot b_1)$	Note $(b)_2$	Points $(a \cdot b_2)$	Note $(b)_3$	Points $(a \cdot b_3)$
First							
Second							
Etc...							
sum	-	-		-		-	

15.2.5 RISK ASSESSMENT

Once the best alternatives have been chosen, we'll move on to checking the risks, or adverse consequences, that could result from the chosen alternatives.

The degree of probability of each of the consequences occurring and the severity of the consequences should they occur should be defined (Tabela 67):

PROBABILITY

- (A) High: will happen;
- (M) Average: it can happen;
- (B) Low: unlikely to happen.

GRAVITY

- (A) High: very important;
- (M) Average: important;
- (B) Low: negligible.

Tabela 67 - Assessing the risks of the decision

Risks (adverse consequences)	probability			gravity		
	A	M	B	A	M	B
First adverse consequence						
Second adverse consequence						
Etc...						

15.2.6 DEGREE OF COMFORT WITH DECISIONS

Once the risks have been defined, we must assess the degree of comfort for decision-making, according to probability and severity, using the Tabela 68 .

Tabela 68 - Degree of comfort with decisions

Degree of comfort		Probability	Gravity
I have no comfort	↓	High	High
I have no comfort	↓	High	Average
I'm comfortable	↑	High	Low
Watch out	↔	Average	High
Watch out	↔	Average	Average
I'm comfortable	↑	Average	Low
I'm comfortable	↑	Low	High
I'm comfortable	↑	Low	Average
I'm comfortable	↑	Low	Low

Depending on the risks, it is possible that none of the alternatives chosen will give us the comfort to decide, so we may have to abandon the alternatives we chose initially and start considering 3ª , or 4ª , or even another with a lower score.

Finally, once we have chosen the alternative that most satisfies our selection criteria and gives us the greatest comfort in deciding, we can move on to planning actions to deal with the adverse consequences, which is our last stage of decision-making.

In cases of time pressure, we can summarize decision-making in just four steps: 1) Purpose of the decision, 2) Mandatory objectives, 3) Alternatives and 4) Adverse consequences.

15.2.7 Actions on risks

We should always consider taking action on the risks in order to eliminate discomfort about the decisions we make.

We can decide to live with the risks if they are very remote and of low severity, but in other cases we can decide to minimize or even eliminate them.

There are two types of action to minimize or eliminate risks:

- Protection actions;
- Prevention actions.

Protection actions

These are the actions we take on the effect of the risk in order to reduce its severity if it happens.

PREVENTION ACTIONS

These are the actions we take against causes in order to reduce the likelihood of them happening.

15.3 Methodology for operational monitoring systems

The starting point for the development of a decision monitoring system should be the identification of a situation that is currently tending to worsen and that has periodic falls in its indices, in other words: situations that are cyclical problems. The prior existence of data from a simple monitoring system can be the starting point for the study. The second step in the development is to assess the extent of the losses that are occurring, or have occurred, in the activity that has problems. Finally, we need to determine the time needed to solve the problem in each of the intensities in which it occurs. Once this is done, we can determine the critical points and the decisions to be taken in each case, based on our experience or the data available.

For a system of this type to be developed, we must always have a lot of data, people who have extensive experience in the subject and meet the following conditions:

1) The situation is a problem and deserves attention;
2) The cause of the problem must already have been identified by any problem analysis method;
3) The solutions to the problem have already been identified in their most diverse variations;
4) The problem is cyclical, i.e. you can't definitively eliminate the causes, you can only control the causes and effects.

If the causes of the problem are not known or if the solutions have not yet been determined, deal with this before starting to develop decision monitoring. If the problem is not cyclical, then there is no need for decision monitoring and we should review the case, as there is a high probability that it will be possible to eliminate the causes and then eliminate the problem once and for all.

Measuring effects, or results, provides us with valuable data for identifying activities that have problems. All available data should be collected and people involved in the activity who can contribute information should be identified.

We're going to analyze three cases based on situations that are problems for many companies:

1° Case - Damage caused by leafcutter ants in reforestation: In this specific case, the cause of the problem is known: leaf-cutting ants; and so is the solution: combating the ants with formicides. This is a typical case for developing a decision monitoring system. Companies usually have people with extensive experience in identifying areas that need to be combated, characterizing the intensity of the attack and assessing the damage caused, which can compensate for the lack of records on these variables. It is a good idea to check how much is spent on formicides, labor and equipment to combat them, as well as estimating the damage caused in terms of a reduction in the annual increase in wood per unit area. A reforestation attacked by ants always tends to increase its losses over time, in other words, it tends to get worse. However, we know in advance that the solution is not permanent; some time after the ant killer is applied, the ants return to the area and the problem becomes cyclical.

2º Case - Reduction in tree growth due to weed competition: Here, too, the cause is known: weed competition; and the solution: chemical or conventional weeding. Weeding isn't a definitive solution either, as the weeds re-establish and develop over time. The problem is cyclical, the cause cannot be eliminated, only controlled.

3º Case - Losses caused by road accidents: In this last situation, the cause is not known and needs to be studied, which can be done using data from internal company surveys, or from the federal highway police, which keeps regular statistics on the causes of accidents. With this data, a Pareto diagram can be drawn up to identify the most frequent cause(s). To illustrate, see the example in Figura 53 with fictitious data:

Acidentes na BR-000 em 1997		
CAUSAS	Nº	% ACUM.
Fator Humano	500	50%
Condições do Veículo	250	75%
Condições da Sinalização	167	92%
Condições da Rodovia	83	100%
Total	1.000	100%

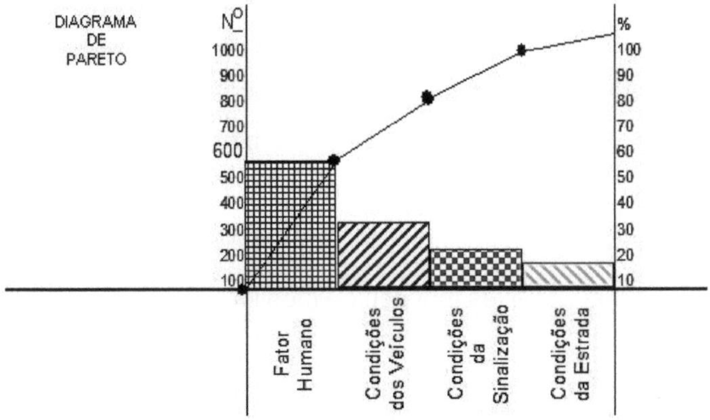

Figura 53 - Pareto diagram of the causes of road accidents.

In this example, it can be seen that the human factor is responsible for 50% of the causes, which added to the condition of the vehicles, comes to 75%. We recommend working primarily with the human factor to reduce accidents. But if we want to achieve better results, we need to include vehicle condition control in our actions. The human factor can be reduced with training, but there is a need for retraining, because what is absorbed in training is gradually forgotten and left unused, making the problem cyclical. Also, vehicle maintenance, which is the solution to the second cause, needs to be regular, as parts wear out continuously.

As we have seen, all three cases meet the conditions for being the subject of decision monitoring and we can move forward with the situations.

Another way of identifying cyclical problems in companies is to examine their accounting reports. Examining balance sheets reveals where money is spent, and where there are large ongoing expenses, there may also be large cyclical problems. Looking for cyclical problems that deserve attention in companies and monitoring them is a very effective way of reducing costs and improving quality.

Other activities that can be monitored are:

- Machinery, vehicles and equipment, with the aim of reducing refurbishment and maintenance costs, based on the value of the intervention in relation to that of the equipment;
- Roads, with the aim of reducing maintenance and refurbishment costs in relation to the value of the end product or transport service being provided via the road in question. It must consider its use or vehicle flow as a criterion to characterize its importance;
- Environment, to control environmental damage caused by industries, forestry companies, transport companies, etc;
- Services provided by third parties to control production (quality and quantity), avoiding errors and delays;
- Stocks, to be reduced to a level compatible with needs and at minimum cost.

Identifying a problem, identifying its cause, discovering the solution and knowing that the problem is cyclical, is the key to realizing the need and creating a monitoring system for continuous improvement.

First of all, information must be gathered . First of all, it is necessary to identify the people involved in the problem and gather all the information about it, by reviewing the literature and consulting people, covering its causes, effects, solution, frequency and costs. You can hold a meeting with the people involved and ask them to gather all the information they have; you should check what information is needed to complement what is available and ask someone present at the meeting to collect it. You can then move on to PDCA, i.e. the four sequential steps: plan, execute, control and correct:

1ª STAGE - PLANNING

In the planning stage, you should:

- Define the purpose of the monitoring;
- Determine the causes to be controlled and their variables;
- Define how the measurements or surveys will be carried out;
- Determine the frequency of the surveys;
- Create decision criteria;
- Define the actions to be taken in each situation using formulas, flowcharts or tables, depending on the decision criteria.

2ª STAGE - EXECUTION

In the execution stage, you must:

- Train all staff in the execution of activities and surveys, in the use of flowcharts, tables and decision formulas and in the execution of selected actions;

- Carry out scheduled surveys;
- Determine the actions to be carried out according to formulas, flowcharts or tables, in accordance with what has been planned;
- Execute the selected actions;

3ª Stage - Control

At this stage, the effectiveness of the actions taken must be assessed, i.e. whether the actions controlled the effects and whether the planned survey and methodology led to the correct decision.

The ineffectiveness of the actions may be due to the following causes:

- The action was poorly executed (by mistake or delay);
- The correct action was not chosen, which can happen for three reasons: 1) the decision methodology is wrong; 2) there was a failure in the execution of the survey; 3) there was a failure in the planning of the survey.

Once the cause of the ineffectiveness of the action taken has been found, we move on to the next stage.

4ª Stage - Corrections

In the event that the ineffectiveness of the action was due to its incorrect execution, corrective action must be taken with regard to the execution of the action and the people responsible must be trained in the new form of execution.

If it is found that the actions carried out have not been effective due to the inadequacy of the action, the necessary corrections to the methodology should be determined so that decision-making is effective next time, and the staff should be trained in the changes made.

Note that the corrections stage is followed by a new planning stage, where we have new information represented by the first planning stage and the defects, or problems, found when it was carried out, it has already been determined what needs to be done to correct the defects and it is necessary to carry out a new planning stage with this data, in other words, it is the start of a new cycle. Then it's back to the execution stage, evaluation, correction and so on.

The following three examples are factual, they were actually developed in forestry companies for: monitoring leaf-cutting ants, monitoring matocompetition and monitoring wood transportation.

15.4 Monitoring leaf-cutting ants

This is a real case developed for a large *Eucalyptus* pulp industry in 1993. We have tried to demonstrate all the stages of the development of this example, so that the process can be understood from start to finish.

The system was developed in 1992 and is applicable to *Eucalyptus* forest stands over six months old when the greatest risk is represented by saúvas, in the north of the state of Espírito Santo and the southeast of Bahia. It can be adapted to other regions using the methodology described here;

The productivity of one person with a vehicle, trained to carry out a field survey of leaf-cutting ants, is 500 hectares per day, under the conditions in which it was carried out.

The system led to an immediate reduction of around 30% in the cost of fighting ants.

The stages of the development process were similar to those of the PDCA. We started by gathering information and then followed the four stages: planning; execution; control; correction.

INFORMATION GATHERING

The existing information referred to the costs of the activity, which amounted to R$0.25/m^3 at the end of the *Eucalyptus* plantation rotation. This is a considerable amount if we take into account that the company produced around 2.2 million m^3 /year, representing a total annual cost of R$ 550 thousand. This already demonstrates the importance of the issue, but we still have to add up the losses caused by ant damage. It is estimated that an anthill with an area of 10 m^2 is capable of killing a 42-month-old *Eucalyptus* plant in a period of four months (the average age of the stands). Assuming an average annual increase of 33m^3 /ha and 1000 plants/ha, we would have lost 0.1155 m^3 in the four months of attack, or 0.3465m^3 /year per anthill with an area of 10 m^2 . Multiply this figure by the number of anthills found throughout the company's forest area and you have an estimate of the total damage caused by anthills of this size.

First of all, all the staff involved in controlling and combating leaf-cutting ants, from foremen to technicians and engineers, met in a room. Through brainstorming, it was discovered that the main indicators of risk in relation to leaf-cutting ants, or variables of causes, were as follows:

- Size of anthills: the bigger the anthills, the more risk they pose;
- Number of anthills: the more anthills, the more damage they cause;
- Percentage of damage caused to the tree crown: the more the tree is attacked, the more it slows down its growth until it dies;
- Number of trees attacked: the more trees attacked, the greater the severity of the losses;
- Concentration points: according to the technicians, heavily infested areas are problematic, because as the canopy shrinks, light penetrates into the interior of the forest, encouraging the growth of the understorey, which will compete with the trees;
- Planting age: the younger the trees, the more susceptible they are and the less resistance they offer to attack.

With this in hand, a survey was carried out in several pilot areas up to three years old, each covering around 100 hectares, for a total of around 8,000 hectares.

1ᴬ STAGE - PLANNING

A statistical analysis of the pilot survey data was carried out. The staff met again. The results of the surveys were presented and then the question arose: At what size do anthills pose the most serious risk and for what size trees? At this point, we started distributing the data into classes. We then built a double entry table by anthill size class and by tree age class with the critical anthill size values for each age class. The second question that arose helped to define the real risk of anthills by size: How much damage was caused by each class of anthill, in each age class of trees. This was an intermediate phase of studying the data in order to better understand the problem.

The aim of this monitoring is to decide whether there is a need to combat with formicides and how much time is left to carry it out without significant losses. Based on this, we decided that the main indicator was the deadline for carrying out the fight and tried to answer the following questions in relation to our variables:

- What is the deadline for combating an anthill before it causes significant damage in each class of anthill size up to 1 m^2 , 2 m^2 , 3 m^2 , 4 m^2 , 5 m^2 , 10 m^2 , 20 m^2 , 30 m^2 , 40 m^2 and 50 m^2 ? (anthills of 50 or more m^2 of area are considered to present maximum risk[3]);
- How long does it take to combat an area with 10, 20, 30, 40, 50, 100, 150 and 200 anthills per hectare? (200 anthills/ha was considered the maximum risk);
- What is the deadline for combating an area in which the percentage of damage to the tree crowns is 5%, 10%, 15%, 20%, 25%, 30%, 35%, 40%, 45% and 50% of their height (above 50% they were considered irrecoverable)?
- What is the deadline for combating ants in an area where, in the general survey, up to 5, 10, 15, 20, 25, 30, 35, 40, 45 and 50 trees were found to be attacked? (We believe that in an area of 100 ha,

[3] Maximum risk = a situation involving major losses and requiring immediate ant control.

there would hardly be more than 50 trees attacked and that this number would be the maximum risk);

- What is the deadline for combating an area with 1, 2 and 3 concentration points? (From the experience of the staff, it was considered that 3 concentration points would be the maximum risk);
- What is the timeframe for combating an area with average infestation (referring to the pilot survey and the experience of each of the participants) and with ages of 1/2, 1, 2, 3, 4, 5, 6 and 7 years? (As the stands were subjected to systematic combating in predetermined periods up to 6 months of age, no stands were sampled up to this age).

Based on the answers to these questions, a time scale was set up every 15 days, starting with 0 up to 360 days, and the sum of the values found for each possible combination was converted into a risk factor. The final scale, developed after the evaluation phase, is shown in Tabela 70 .

Below we describe the six indicators, or variables, considered for the development of the monitoring system for leaf-cutting ants: Size of anthills, number of anthills per hectare, Percentage of canopies with damage, concentration points and age of the plantation.

Size of anthills (T) - in order to measure the surface area of an anthill, it is necessary to locate all its scouts, draw a line in the direction of the greatest distance between two scouts representing its length (c) and another, transversal to the first, representing its maximum width (L), as in Figura 54 The surface area or size of the anthill (T) is given by the product of its width and length.

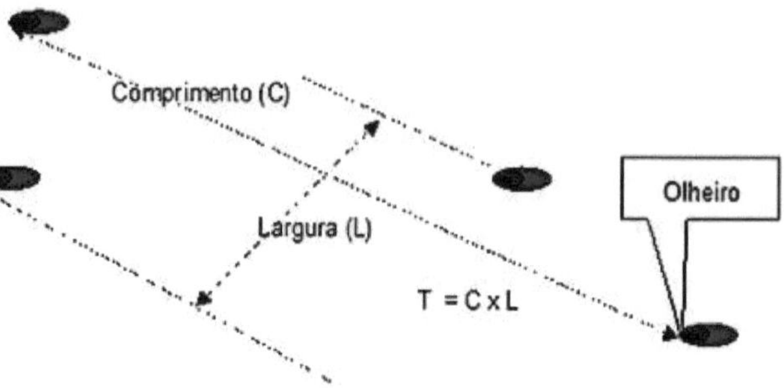

Figura 54 - Measuring an anthill.

Number of anthills per hectare - to determine the number of anthills per hectare, between 2 and 6% of the forest area is sampled, in strips distributed in the places with the highest ant infestation, where the anthills found in the strip between four rows of planted trees are counted, as in Figura 55 .

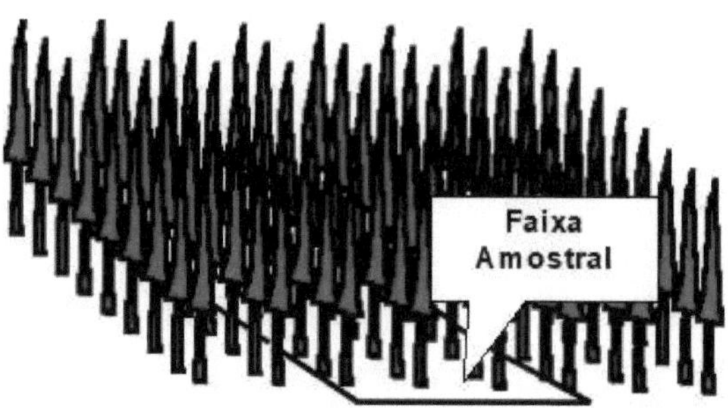

Figura 55 - Sampling range for monitoring leaf-cutting ants.

Count the anthills found in the sampling strip and divide the number found by the area of the strip, which is given by 4 times the distance between two planting lines, multiplied by the length of the strip.

Percentage of crowns with damage - this is the percentage of the height, with damage, of the crown of the tree that was most attacked by ants; see example estimate in Figura 56 For this assessment, the most attacked tree in the forest stand being monitored is deliberately chosen.

Nº of cut trees - this is the total number of all damaged trees found in the stand being monitored, regardless of whether or not they are in the sampling bands.

Concentration point - is an area with a radius of up to 50 m, where at least three anthills occur and whose total area of anthills is greater than 25 m .[2]

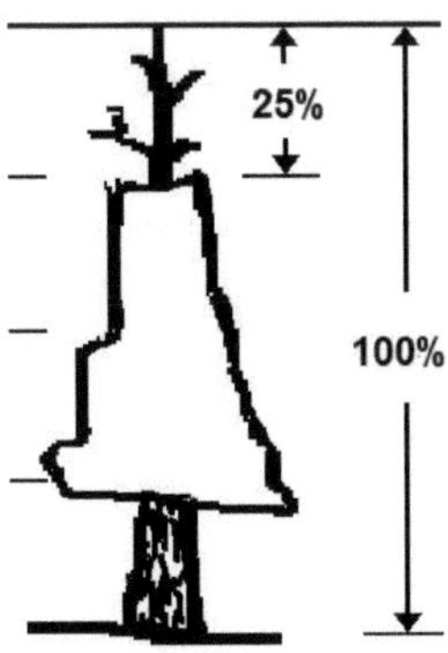

Figura 56 - Percentage of tree height with damage .

Planting age - the age of the forest stand expressed in months.

For data collection, the following table was developed Tabela 69 was developed, showing a hypothetical example of a survey in one area. The table is suitable for monitoring areas of mixed forest with a surface area of around 100 hectares.

2A STAGE - EXECUTION

In this case, in particular, the survey is carried out by always looking for the worst, i.e. all the roads in the area to be surveyed are traveled and sampling is carried out in strips that are located at the points of greatest infestation, looking for the largest anthill, the tree with the most damage to the crown, counting all the trees cut down by ants, identifying all the concentration points and noting everything on the survey spreadsheet (Tabela 69).

Tabela 69 - Leafcutter ant survey form

Location:							Planting:		Last fight:		Date:	
Project:				Plot:			__/__/__		__/__/__		__/__/__	
Risk factor	**Size of anthills (m^2)**		**No. of anthills per ha**		**% of crowns with damage**		**Number of trees cut down**		**No. of of concentration**		**Planting age (months)**	
1	-		≤ 10		-		-		0	1	≥ 72	
2	≤ 1		11-20	2	-		-		-		60 - 71	
3	1,1 - 2,0		21 - 30		< 5		< 5		-		48 - 59	
8	2,1 - 3,0	8	31 - 50		6 - 10		6 - 10	8	-		24 - 47	
10	3,1 - 4,0		51-100		11 - 15		11 - 15		-		12 - 23	10
18	4,1 - 10,0		101- 150		16 - 20	18	16 - 20		-		10 - 11	
21	10,1-30,0		151-200		21 - 30		21 - 30		1		06 - 09	
24	> 30		> 200		≥ 30		≥ 30		≥ 2		≤ 6	
Risk	=	8	+	2	+	18	+	8	+	1	+	10

The table is annotated with the number corresponding to the occurrence factor. For example, the largest anthill observed was around 2.5 m^2 , belonging to the size class

>2.0m and$^2 \leq 3.0m^2$, which corresponds to factor 8 and this value is noted in the column corresponding to the size of the anthills; the sampling revealed 17 anthills/ha, factor 2; the tree with the most damage had 25% of its crown affected, factor 18; a total of 15 trees were attacked, factor 8; there were no concentration points, factor 1; the plantation was 28 months old, factor 10. Using this data, the risk index was calculated (Tabela 70) and the action to be taken was determined.

To calculate the risk index, the points found are added together. The risk index calculated with the example in Tabela 69 is:

Risk index = 8 + 2 + 18 + 8 + 1 + 10 = ***47***

In the case studied, the risk index = 47 corresponds to medium risk in the Tabela 70 which implies immediate action to combat leaf-cutting ants in the sampled stand.

At this stage, a new survey was carried out of the same areas where the pilot survey was carried out and the actions determined by the calculated risk were carried out.

Tabela 70 - Actions to take depending on the risk (decision criteria)

Contents	Risk	Action	Deadline
0 - 10	None	Monitoring	12 months
11 - 25	Very small	Monitoring	6 months
26 - 45	Small	Monitoring	3 months
46 - 75	Medium	Combat	1 month
76 - 100	High	Combat	15 days
>100	Very high	Combat	Immediate

3ª Stage - Control

After a month, a new survey was carried out in the areas where the monitoring had been determined to be combated, to assess the results of this action. The vast majority of cases were found to be effective. In the cases where this was not the case, the action had been determined efficiently, but its execution had been deficient. The corrections made in these cases came down to better training the personnel involved in carrying them out.

4ª Stage - Corrections

With the data from the second survey, the table of actions according to risk was corrected. The only corrections needed were to redistribute classes and revise the original survey table, reducing the information collected. The original table involved more calculations, which were eliminated. It is true that over time, due to the influence of biological and climatic factors, there will be a need for corrections to the system, but it has been tested on a large scale and works satisfactorily under the conditions in which it was developed.

15.5 Weed monitoring

The following example is hypothetical and will only serve as an example, and should be adapted to real situations. We developed a similar system for a pulp mill in 1993, which was simpler and more dependent on supervisory decisions. The model presented here is more complex and automatic, but has not been tested in the field to evaluate the results.

Forestry companies have to spend a considerable amount of money weeding their *Eucalyptus* plantations. In addition, the herbicides used are environmentally damaging and conventional weeding (mechanical and/or manual) destroys the structure of the topsoil and harms the microorganisms in this layer by disturbing and exposing the soil.

INFORMATION GATHERING

Activity costs at the time: R$0.35/m^3 until the wood is harvested at seven years of age. Cause and effect study: some of the causes of the reduction in *Eucalyptus* growth are mentioned in the diagram in Figura 57 .

The main weeds occurring in the region (variables of causes) are:

- broadleaves resistant to the herbicide: tansy *(Laportea aestuans*), castor bean (*Ricinus communis*), guanxuma (*Sida spp.*);
- narrow-leaved grasses susceptible to the herbicide: colonião (*Panicum maximum*), *brachiaria* (*Brachiaria spp.*), quicuio (*Urochloa humidicula*), pernambuco grass (*Paspalum notatum*).

The cause considered in this study was matocompetition (weeds) and the effect studied was the reduction in the growth of the *Eucalyptus*, which is to be avoided.

Solutions to the problem: the solution to the occurrence of weeds that compete with the *Eucalyptus* is to carry out chemical or conventional weeding (mechanical and/or manual), depending on the type of weed that occurs, its size in relation to the *Eucalyptus* and the size of the latter. When the *Eucalyptus* plants are up to 1.5 meters high, chemical weeding should not be carried out to avoid damaging the *Eucalyptus* itself with the herbicide.

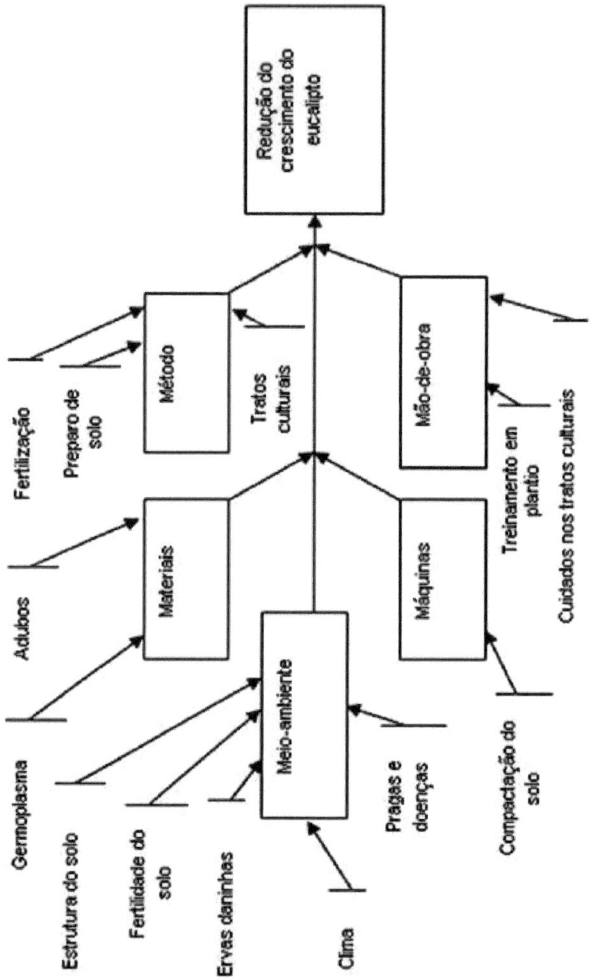

Figura 57 - Cause and Effect Diagram

When broadleaves are more prevalent, accounting for more than 65% of the total weed cover, weeding should be conventional, as the herbicide used for grasses has little effect on broadleaves and the herbicide used for broadleaves is not only harmful to *Eucalyptus*, but also very expensive.

Stage 1 - Planning

At this stage we must define the objective of the monitoring, which is to prevent weeds from competing with the *Eucalyptus*. This means that the mere presence of weeds does not mean that they are in competition with the crop. So the question is to determine when the weeds are in competition, or are at risk of doing so within a certain period of time, when a new survey should be carried out, or weeding itself as a control measure. In Tabela 71 and Tabela 72 show the decision criteria for eucalyptus plantations up to 5 m high, drawn up using the empirical knowledge of the team of forest supervisors. The action codes in Tabela 71 e Tabela 72 are as follows:

- M 60 - Monitor after 60 days;
- M 50 - Monitor after 50 days;
- M 40 - Monitor after 40 days;
- M 30 - Monitor after 30 days;
- C 15 - Weeding after 15 days;
- C 7 - Weed after 1 week;
- C - Weeding immediately;
- CP - Partial weeding (only of the infested area).

Tabela 71 - Solutions for broad leaves

Height of herbs in relation to *Eucalyptus*	Covering the soil surface with weeds						
	10%	**20%**	**30%**	**40%**	**50%**	**60%**	**70%**
10%	M 60	M 60	M 60	M 50	M 40	M 30	M 20
20%	M 60	M 60	M 50	M 40	M 30	M 20	C 15
30%	M 60	M 50	M 40	M 30	M 20	C 15	C 7
40%	M 50	M 40	M 30	M 20	C 15	C 7	C
50%	M 40	M 30	M 20	C 15	C 7	C	C
60%	M 20	M 20	C 15	C 7	C	C	C
70%	CP	CP	C 7	C	C	C	C

Tabela 72 - Solutions for grasses

Height of herbs in relation to *Eucalyptus*	Covering the soil surface with weeds						
	10 %	**20%**	**30%**	**40%**	**50%**	**60%**	**70%**
5%	M 60	M 60	M 50	M 40	M 30	M 20	C 15
10%	M 60	M 50	M 40	M 30	M 20	C 15	C 7
15%	M 50	M 40	M 30	M 20	C 15	C 7	C
20%	M 40	M 30	M 20	C 15	C 7	C	C
25%	M 30	M 30	C 15	C 7	C	C	C
30%	M 30	M 20	C 7	C	C	C	C
35%	M 20	M 20	C	C 7	C	C	C
40%	M 20	M 20	C	C	C	C	C
50%	M 20	CP	C	C	C 7	C	C
60%	CP	CP	C	C	C	C	C
70%	CP	CP	C	C	C	C	C

If the decision from the previous tables is to weed, the type of weeding will be decided using the flowchart in Figura 58 .

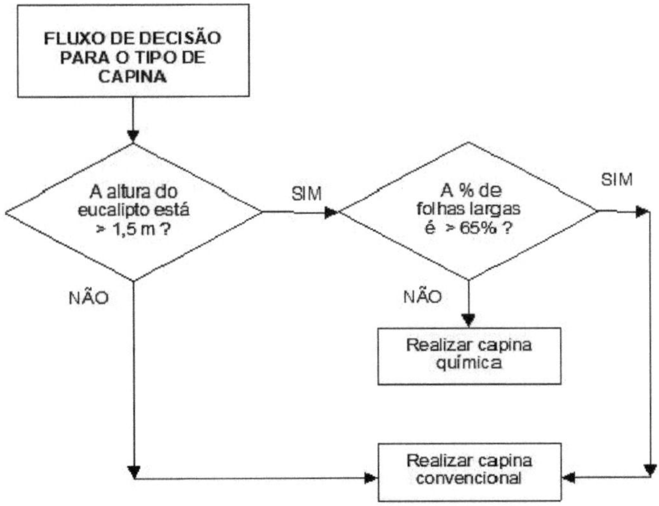

Figura 58 - Flowchart for deciding on the type of weeding .

The survey form developed for monitoring matocompetition is shown in Tabela 73 .

Chemical weeding should be carried out with a herbicide suitable for grasses, which can be applied using the most effective equipment for the situation, or whatever is available. Conventional weeding can be carried out with mechanical weeding between the planting rows, complemented by manual weeding in the row.

The survey table should contain data on the location of the area, the date, the occurring herb species and their proportional frequency, the size of the *Eucalyptus* and the proportional size of the occurring species, as in Tabela 73 below.

Tabela 73 - Spreadsheet for monitoring matocompetition.

Location:	*Eucalyptus* height:	Last weeding:	Date:
	m.	__/__/__	__/__/__

Project:	Plot:	Note

Occurring herbs		Proportional frequency of herbs										Height relative to *Eucalyptus*									
		10%	20%	30%	40%	50%	60%	70%	80%	90%	100%	10%	20%	30%	40%	50%	60%	70%	80%	90%	>90%
Wide leaves	Cansanção																				
	Castor beans																				
	Guanxuma																				
Subtotal F. Largas												F. Lar. coverage (%):									
Small Grasses	Brachiaria																				
	Quicuio																				
	Penambuco																				
Subtotal G. Small												Coverage of small children (%):									
Large grams	Settler																				
Subtotal G. Large												G. Gm. coverage (%):									
Total												% grass cover:									

Notes: Proportional frequency of weeds is the percentage with which one weed occurs in relation to the others, considering the total infestation as 100%. Percentage of cover with weeds is the percentage of the total soil surface of the assessed area that is infested with weeds (it should be equal to the sum of the cover of broadleaves, small grasses and large grasses).

The dose of herbicide must be determined by the technician, depending on the vital stage of the most common weeds in the infested area. This decision has not been automated due to its complexity.

At this point we have completed the planning stage and can move on to the next one.

2ª Stage - Execution

At this stage, we would have to carry out a series of surveys, decide on the necessary actions as planned and carry them out. Then we would move on to the next stage.

Stage 3 - Control

Once the effect can be seen in relation to the weeding carried out, it is necessary to check the result of the weeding. There are two possibilities: the weeding had the desired effect, or it didn't (the desired effect, in this case, is the elimination of weed competition with the *Eucalyptus*). If the effect was not as desired, the reason or causes for the ineffectiveness of the actions or weeding must be verified.

Stage 4 - Corrections

Once the causes for the ineffectiveness of the actions carried out have been discovered, which were determined according to the survey and decision methodology defined in the planning, the necessary corrective actions must be determined and new planning started, restarting the cycle.

Note: In any case, this methodology should be adjusted, as it was designed with values based on the empirical knowledge of a team of forestry technicians and engineers, but the model has not been adjusted with field tests.

15.6 Monitoring timber transportation

The example we are about to present was developed for a transport company in the state of Espírito Santo in 1994. The company transported wood for a large pulp industry. At the time, it had 45 mechanical horses transporting wood, working 24 hours a day, and had an average of two accidents a month. Based on the assumption that the main cause of accidents was due to human factors, in addition to measures in other areas, in our first contacts, the company hired a defensive driving course for the drivers. In our opinion, the course alone would have had a great effect, but it represented only one alternative solution to the problem, and something more complete was proposed. The first step was to train transport monitors, for whom a place was obtained on the Mercedes Transport Monitors Training course, training instructors at SENAC (National Trade Service) and courses were given in defensive driving, economical driving of timber transport vehicles and transport monitoring at the company itself, in order to prepare them to train the drivers and carry out the monitoring. The survey form developed is shown in Tabela 80 . Before the system was implemented, as has already been said, there were two accidents a month, with all the consequences: people injured, loss of income, costs of repairing the vehicle, costs of four drivers stopped per vehicle, etc.

It can be seen that the personal factor constitutes a cause in method (defensive driving) and in manpower (knowledge, physical conditioning and experience). The analysis of causes and effects was carried out as shown in Figura 59 .

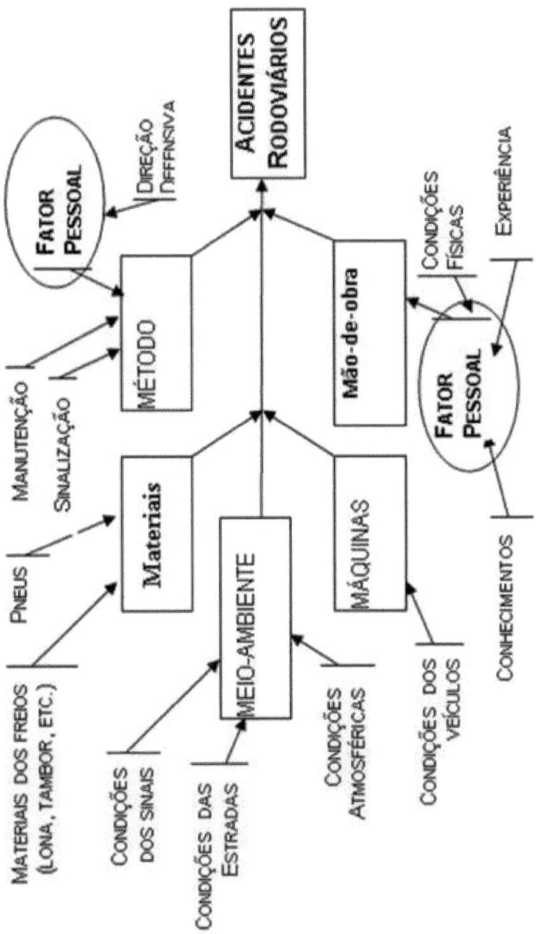

Figura 59 - Cause and effect diagram for road accidents.

The priority has already been established: the main cause is the human factor. We now need to see what its variations are, or separate the parts. What do we mean when we say that the human factor is the main cause of accidents? We mean that human beings make a lot of mistakes when driving. And what else? Well, the answer to this question is that we make mistakes when using a vehicle in the situations in the Tabela 74 .

Tabela 74 - When mistakes are made and in what situations.

Situations in which a mistake is made	Care to be taken
Overtake; Passing at intersections; Going around bends, up and down hills; Crossing bridges and viaducts; Brake and pull over; Keep your distance from other vehicles; Use the headlights and signal with the vehicle's lights; Drive at an appropriate speed; Driving in urban and industrial areas;	Taking care of the load (lashing); Wear the vehicle's protective equipment; Keeping track of supplies and maintaining vehicle hygiene. Inspect the vehicle periodically; Check the tires of cargo vehicles every time we stop; Constantly check the functioning of the electrical system when driving on uneven roads.

All the variables mentioned above are causes of accidents, or aggravating factors in their consequences. But the last question to separate all the parts of the problem remains: What leads us to say that these are the variables that cause accidents in terms of the human factor? Well, in this case the answer is: mainly accident reports and our experience.

We have completed the information-gathering stage, already knowing the situation and the problem well, and we have moved on to the stage of deciding what to do and, with that, to the planning stage.

1A STAGE - PLANNING

The purpose of the decision in this case is to reduce road accident losses caused by the human factor. The decision criteria are shown in Tabela 75 .

Tabela 75 - Decision Criteria

Mandatory criteria:	Desirable criteria:
Reduce the number of accidents;	Improving the working environment;
Increase the quality of service;	Low implementation costs;
Improving drivers' fitness for the job;	Increase knowledge of economic management techniques;
Increase knowledge of defensive driving techniques.	Reduce vehicle maintenance costs.

The alternatives for solutions to reduce road accidents raised by brainstorming were:

- Implement a rigorous inspection system;
- Create standards for timber transportation services;
- Bring staff together and get more commitment and responsibility when driving vehicles (commitment);
- Training drivers in defensive driving;

- Implement a punishment system for drivers who fail to comply;
- Implement an evaluation system and rewards for good performance;
- Implement a work quality monitoring system;
- Adopt the use of breathalyzers;
- The use of the tachograph was not considered because it was adopted definitively.

The alternatives are compared against the decision criteria as shown in Figura 60 .

Critérios Obrigatórios		Fiscalização rigorosa		Normalização		Obter compromisso		Curso de Direção Defensiva		Punição		Prêmios		Monitoramento da qualidade		Bafômetro	
(Descrição)		Atende	Não	Atende	Não	Atende	Não	Atende	Não	Atende	Não	Atende	Não	Atende	Não	Atende	Não
< N° de Acidentes		◆		◆		◆		◆		◆		◆		◆		◆	
> qualidade do serviço		◆		◆		◆		◆		◆		◆		◆		◆	
> condições físicas		◆		◆		◆		◆		◆		◆		◆		◆	
Uso da direção defensiva		◆		◆		◆		◆		◆		◆		◆		◆	
Critérios Desejáveis	A	B_1	A XB_1	B_2	A XB_2	B_3	A XB_3	B_4	A XB_4	B_5	A XB_5	B_6	A XB_6	B_7	A XB_7	B_8	A XB_8
(Descrição)	Peso	Nota	Pontos	Nota	Pontos	Nota	Pontos	Nota	Pontos	Nota	Pontos	Nota	Pontos	Nota	Pontos	Nota	Pontos
Melhorar ambiente	1	0		1	1	1	1	1	1	0		0		1	1	0	
Barato implantar	2	0	2	0		1	2	1	2	1	2	0		0		1	2
Condução econômica	2	0		1	2	1	2	1	2	0		1	2	1	2	0	
< manutenção	2	1		1	2	1	2	1	2	0		1	2	1	2	0	
Total	x	x	**2**	x	**5**	X	**7**	x	**7**	x	**2**	x	**4**	x	**5**	x	**2**

Figura 60 - Comparison of alternatives with mandatory and desirable criteria.

As you can see, all the proposed alternatives meet the mandatory criteria and none have been eliminated so far. With regard to the desirable criteria, four alternatives scored five points or more and were chosen for implementation, namely: standardizing the timber transport service; bringing together the timber transport staff and getting them involved in the whole process to commit them; training drivers in defensive driving and economical driving of cargo transport vehicles; and implementing a system for monitoring the quality of work in the timber transport service. This was the final proposal presented to the company, which was accepted and implemented.

The risks of the decisions were assessed according to Tabela 76 .

Tabela 76 - Risk assessment of decisions

Risks (Adverse Consequences)	Probability			Gravity		
	High	Average	Low	High	Average	Low
Alternative: Standardization	Inspire Care					
Doing it wrong		M		A		
It takes a long time to do		M			M	
Can't do it			B	A		
Alternative: Get Commitment	Comfortable					
Doing it wrong			B	A		
It takes a long time to do			B		M	
Can't do it			B	A		
Alternative: Defensive Driving Course	Inspire Care					
Doing it wrong			B	A		
It takes a long time to do		M			M	
Can't do it			B	A		
Alternative: Quality monitoring	Comfortable					
Doing it wrong			B	A		
It takes a long time to do		B			M	
Can't do it			B		M	

Where A = high; M = medium; B = low.

In none of the cases were the alternatives selected considered difficult to implement, but in two cases their implementation required special care that was taken into account.

The surveys were scheduled to be carried out weekly using the Tabela 77 the monitors were trained and the selected alternatives were started.

Tabela 77 - Traffic monitoring

Date:	Correct		Incorrect	
Items observed	**Counting**	**Points**	**Counting**	**Points**
Seat belt				
Overtaking				
Driving on bends/aclives/declives				
Driving on bridges/viaducts				
Braking/tailgating maneuvers				
Keeping your distance from the vehicle in front				
Use of headlights and flashers				
Traffic in an industrial area				
Traffic in urban areas				
Supply monitoring				
Load lashing				
Vehicle inspection				
Electricity system situation				
Tire situation				
Internal/external hygiene of the vehicle				
GRAND TOTAL OF POINTS	Correct		Incorrect	
	%		%	

Comments:

2ᴬ STAGE - EXECUTION

At first, mediocre rates of 65% of correct points were obtained, but with the identification of the causes of non-compliance with the requirements of the survey table and the adoption of the necessary corrections in consensus meetings with the defaulters themselves, in about 4 months there was an improvement of over 90% and for several weeks 100% was achieved, with periods of more than 60 days without accidents.

3ª STAGE - CONTROL

In the planned system, any error observed in the surveys is analyzed together with the person who committed it, and the occurrence is taken to training sessions with the drivers, without identifying the person. The aim of the meetings is to find solutions to the causes of the errors, which are often futile or purely inattentive, but with the involvement of the offender in the decision-making process, their commitment is practically total.

4ª STAGE - CORRECTIONS

Most of the corrections that were necessary concerned the behavior and attention of the drivers, who in some cases were sent for retraining in defensive driving. The system itself has not changed since its implementation.

15.7 Forest monitoring

Forest areas, whether production, planted or natural, whether for preservation or conservation, must be constantly monitored to assess their stock, growth, biodiversity and evolution, in order to make decisions and ensure that their objectives are achieved.

15.7.1 Production areas

Production areas generally undergo three types of forest inventory to be monitored:

- Planting inventory for survival assessment;
- Continuous inventory (or actual monitoring) to classify sites and monitor the growth and evolution of stocks, providing data for forecasting and planning;
- Pre-cut inventory to estimate stocks and production, providing data for harvest planning.

Forest inventories follow methodologies developed for sampling, which are not the subject of this book. Each organization adopts a different type of inventory, usually by management unit, classifying production areas by germplasm, site index and age, within each administrative unit.

In the case of managed native forests, the legislation requires, in addition to phytosociological monitoring, an inventory of 100% of the trees with a diameter above 50 cm or above another diameter value depending on the characteristics of the forests to be managed, and there may be specific state legislation in addition to federal legislation.

15.7.2 Conservation and preservation areas

Conservation and preservation areas, in addition to phytosociological monitoring, also need a periodic fauna inventory, where only the species of animals that occur are identified. Special attention should be paid to the occurrence of threatened or endangered species, both flora and fauna, to ensure that management does not negatively influence their populations and to identify the areas of greatest occurrence, or corridors for species in this condition, which should be preserved, preferably locating legal reserve areas in these spaces.

15.8 Soil monitoring

Soils should have their fertility and structure monitored, preferably but not only before planting or after harvests, in order to identify compaction and the need for nutrient replacement and actions such as furrowing for the new forest, with the aim of maintaining its productive capacity. Soil monitoring and management is the subject of a specific study and is not dealt with in this paper.

15.9 Monitoring forest roads

A road is made up of the carriageway, engineering structures and signs. The quality of each of the three components, generating conservation actions at the right time, is essential to ensure safety, comfort, less wear and tear on vehicles and appropriate speed with shorter travel times, lower road maintenance costs and the safe and comfortable transportation of cargo and people.

After construction, the speed at which roads deteriorate depends mainly on the volume of traffic and the weather. The volume of traffic mainly affects the roadway and the engineering structures, while the weather conditions affect all road components, from the roadway to the drainage systems, engineering structures and signs.

Excellent road planning and construction reduces, but does not prevent, road deterioration. A good monitoring system will ensure the lowest road maintenance and transportation costs while guaranteeing the necessary safety.

Road monitoring involves not only the roadway, but also the structures, drainage systems, horizontal and vertical signage, retaining walls and embankments, lanes, operational buildings and more. Nowadays, high-tech equipment is used to monitor roads, making inspections easier and reducing the time it takes to carry them out. The inspection of culverts can be aided by video, lanes can be inspected by drones and signs can be inspected by retro-reflectance equipment.

The trigger for road maintenance depends mainly on the cost of maintenance itself, the increase in transportation costs due to road deterioration and the increased risk of accidents. Transportation costs are influenced by road conditions in terms of the increased cost of wear and maintenance of vehicles and equipment and increased travel time. All these factors must be monitored so that maintenance can be carried out before the costs increase by more than the cost of the maintenance to be carried out on the road, or accidents occur due to the lack of maintenance of the road and all its components. The main road defects are listed in Tabela 78 .

Tabela 78 - Road surface defects on dirt roads.

Defect	Measure	Severity class			
		0 = Very low	1. Low: not very harmful	2. Average: harmful	3. High: very harmful
Cross section - W	Depth (cm)	< 5	5 - 10	> 1 0	-
Corrugations	Amplitude (cm)	< 2	2 - 5	> 5	-
Open pans	Area (%)	< 5	5 - 20	2 0 - 5 0	> 5 0
Holes	Depth	< 5	5 - 10	1 0 - 2 5	> 2 5
	Maximum (cm)				
Transverse grooves	Depth (cm)	< 2	2 - 5	> 5	-
Longitudinal grooves	Depth (cm)	< 5	5 - 10	1 0 - 2 5	> 2 5
Wheel tracks	Amplitude (cm)	-	5 - 10	1 0 - 2 5	> 2 5
Dusty surface	Area (%)	< 1	2 - 5	5 - 2 0	> 2 0
Settlement / landslide	Different levels (cm)	<5	5 - 10	10 - 25	> 25

Source: Horta (1991).

Eaton et al (1992) carried out a study, concluding on road defects such as: incorrect section, road drainage, corrugations, dust, potholes and loss of aggregate. The authors defined each of these defects as follows.

15.9.1 Improper cross-section

Description

Incorrect cross-section is the result of the road surface not being properly shaped or maintained to carry water to the ditches. This condition is evidenced by water soaking into the road surface, water draining or running along the road surface, lack of a crown on the road or erosion of the road surface caused by water runoff.

Severity levels

- L - Small amounts or evidence of drinking water on the road surface or a completely flat road surface (no cross_slope), or both;
- M - Moderate - Quantities or evidence of water on the lake road surface or a bowl-shaped road surface, or both;
- H - Large amounts or evidence of water on the road surface or severe depressions in the wheel paths on the road surface, or both.

Measurement

Incorrect cross-section is measured in linear feet per 100 feet of section, from the outside of the shoulder to the outside of the shoulder. Different levels of severity can exist within the 100-foot sample unit. A maximum of 100 linear feet can be measured.

15.9.2 Road drainage

Description

Poor drainage causes water to enter gullies. Drainage problems occur when ditches and culverts are not in adequate condition to properly direct and transport runoff water. This condition is evidenced by ditches filled with debris or overflowing with debris, ditches that have not been properly shaped or maintained, water running across or under the road, and areas where ditches have begun to erode into the road.

Severity levels

- L - small amounts of:
  - Moisture or evidence of water in gutters, and
  - Overgrowth or debris in gutters;
- M - moderate amounts of:
  - Puddles of water or evidence of water in gutters,
  - Overgrowth and debris in gutters, and
  - Evidence of erosion in gullies, on embankments or on the road;
- H - large quantities of:
  - Water puddles or evidence of water in gutters,
  - Water running all over the road,
  - Overgrowth and debris in gutters, and
  - Erosion of gutters, embankments or roads.

Measurement

Drainage problems are measured in linear feet per 60 cm section parallel to the centerline of the road, from the break in the outer slope perpendicular to and away from the road. It is possible to have a maximum of 200 linear feet of drainage on the road.

15.9.3 Corrugations

Description

Corrugation, also known as washboarding, is a series of spaced rib-shaped grooves and ditches that occur at reasonably regular intervals. The ridges are perpendicular to the direction of traffic. This type of defect is usually caused by the action of traffic and disintegration of the road surface material. These ridges usually form on grades or curves, in areas of acceleration or deceleration, or in areas where the road is soft or bumpy.

Severity Levels

- L - Corrugations less than 1 inch deep, or of low severity, or both;
- M - Corrugations 1 to 3 inches deep, or medium severity, or both;
- H - Corrugations deeper than 3 inches, high severity, or both.

Measurement

Corrugations are measured in square feet of surface area per 100-foot-long section. It must not exceed the total area of the 10-foot-long section.

15.9.4 Dust

Description

The abrasive action of traffic on unpaved roads eventually loosens the larger aggregate particles of the soil binder. As traffic passes, the dust clouds create a hazard for overtaking or passing vehicles and cause significant environmental problems.

Severity levels

Normal traffic produces the following levels of severity:

- L - Fine dust that does not obstruct visibility;
- M - Moderately thick dust cloud that partially obstructs visibility and causes traffic to slow down; or

- H - A very thick cloud that obstructs visibility and causes traffic to slow down significantly or come to a standstill.

Measurement

The dust is measured by driving a vehicle at about 45 km/h and observing the dust cloud; the dust is estimated as fine, moderately coarse or very coarse. The dust is not classified by density. The severity of the problem is determined by the size of the dust cloud generated by the traffic and the reduction in visibility caused by the dust.

The deduction values for the severity levels are as follows: low - 2 points; medium - 5 points; high - 15 points.

15.9.5 Holes

Description

Potholes are small bowl-shaped depressions in the road surface, which are usually less than 3 feet in diameter. Their growth is accelerated by the accumulation of free moisture inside the hole on rainy days, or even in a dry period by crumbling. Potholes are produced when traffic tears off small pieces of the road surface. The road continues to disintegrate due to loosening of the surface material or weak spots in the base or subgrade.

Severity levels

Severity levels for holes under 3 feet in diameter are based on both the diameter and depth of the hole, according to Tabela 79 .

Tabela 79 - Severity levels for potholes.

Maximum depth	Average diameter			
	< 0,30m	**0.31 to 0.60 m**	**0.61m to 0.91m**	**>0,91m**
1 to 5 cm	Bass	Bass	Medium	Medium
5.1 to 10 cm	Bass	Medium	High	High
> 10 cm	Medium	High	High	High

Eaton et al (1987).

If the hole is more than 3 feet in diameter, the area must be determined in square feet and divided by 5 square feet to find the equivalent number of holes.

Measurement

Potholes are measured by counting the number that are of low, medium and high severity in a 100-foot-long section and recording them separately by severity level.

15.9.6 Grooves

Description

A rut is a shallow depression in the path of the tires. Grooves are caused by permanent deformation in any of the layers of the road or subgrade. It results from repeated traffic loads, especially when the road is soft. Significant rutting can lead to major structural failure of the road.

Severity levels

- L - Grooves less than 1 inch deep, or of low severity, or both;
- M - Furrows 1 to 3 inches deep, or medium-severe, or both;
- H - Furrows deeper than 3 deep, of high severity, or both.

MEASUREMENT

Furrows are measured in square feet of surface area in a 100-foot-long section. The total square feet of furrows must not exceed the total area of the 100-foot-long section.

15.9.7 LOSS OF AGGREGATION

DESCRIPTION

The abrasive action of traffic on unpaved roads eventually loosens the larger aggregate particles from the soil binder. This carries the aggregate base particles away from the road surface or shoulder. Traffic moves disaggregated particles away from the normal path of the tires and forms ridges in the center or along the shoulder of the road or less-traveled area, parallel to the axis of the road.

SEVERITY LEVELS

- L - Disintegrated shoulder or road surface, or a loss of aggregation on the shoulder or less-traveled road surface of less than 2 inches, or both;
- M - Moderate aggregate ridge (2 to 4 inches) on the shoulder or less traveled area of the road, excessively fine and usually found on the road surface;
- H - Wide aggregate ridge (greater than 4 inches) on shoulder or less traveled area of road.

MEASUREMENT

Breakdown is measured in linear feet on a 60 m long section parallel to the center line of the road.

15.9.8 INSPECTION AND ASSESSMENT OF SIDE ROADS

An inspection form for side roads was developed by Cruz (2005) (Figura 61) as described in the work by Eaton et al (1987).

FICHA DE INSPEÇÃO DE ESTRADAS NÃO PAVIMENTADAS		
1. Estrada	2. Trecho:	3. Data:
4. Trechos Selecionados	5. Área do trecho:	6. Avaliador:

7. Desenho	8. Tipos de defeitos 8.1. Seção Transversal Inadequada 8.2. Drenagem Lateral Inadequada 8.3. Corrugações / Ondulações 8.4. Poeira 8.5. Buracos 8.6. Trilhas de Rodas 8.7. Segregação de Agregados

Quantidade e severidade dos defeitos								
Tipos		8.1	8.2	8.3	8.4	8.5	8.6	8.7
Quantidade e Severidade	B							
	M							
	A							

9. Cálculo do URCI

Tipo de defeito	Densidade	Severidade	DV	10. Anotações
Valor de dedução (TDV):		q =	URCI:	Classificação:

Figura 61 - Inspection form for side roads. Source: Cruz (1987).

15.9.9 MONITORING ARTWORK AND SIGNAGE

Construction works, especially bridges, drainage ditches and road culverts, as well as signage, require periodic and permanent monitoring.

Bridges can have problems with their structure, erosion of the wings and the edge of the roadway, especially on dirt roads.

Culverts are more problematic than bridges. In addition to the possibility of erosion on the edges and wings, the loss of material above the culvert can cause cracks and displacement of shackles, which can lead to seepage and external erosion along the shackles. Added to this is the possibility of the culvert becoming blocked due to siltation. It is recommended that a layer of material thicker than 1.5 times the diameter of the culvert is kept above the culvert. Damp or very clayey soil at the base of the culvert can make it easier for the base to deform and cause the shackles to shift, allowing infiltration that can destroy the culvert unnoticed. Careful and regular inspections can highlight the need to repair culverts before the road is interrupted in a rainy season.

Drainage ditches can be silted up, accumulating debris that makes it easier for vegetation to grow, interrupting the flow of water outwards and causing erosion on the sides of the ditch. They can also have cracks that make it easier for water to seep in. When they are only built by opening furrows in the ground, they can erode over time, requiring containment works. Culvert outlet ditches are often built in such a way as to reduce the speed of the water, but when it exits onto the ground, it can erode, gradually undermining the masonry gutter and destroying it.

Signage deteriorates over time and is damaged by passers-by, and must be constantly monitored and replaced to ensure safety on the roads.

15.10 Pest and disease monitoring

Pests and diseases are becoming increasingly widespread in forestry. Their constant monitoring is an unquestionable necessity.

Monitoring can be carried out from the ground, similar to ant monitoring, or from the air and space.

Satellite images make it easier to observe spots on vegetation, the spectrum of which can be correlated with some type of pest or disease. However, the resolution of the images obtained is not sufficient to make categorical statements; on the other hand, it is possible to obtain extremely high-resolution images using drones.

Drones are increasingly being used to monitor cultivated and protected forests. Drones can be used to identify the type of species that is affecting the forest in detail due to the high resolution of the images that can be obtained.

There are times when climatic extremes are more problematic for pests and diseases, while others are not so problematic. The frequency of monitoring should be adjusted for each case as more data and experience is obtained.

15.11 Monitoring forest fires

Traditional fire monitoring is carried out using observation towers with cameras that transmit images in real time. In the past, when there were no digital cameras, this was done by people who stayed at the top of the towers during the hours when fires were most likely to start (usually between 9 am and 6 pm).

Three towers strategically placed at high points in the terrain are needed to cross-reference the angles and determine the coordinates of a fire outbreak. The distance between the towers must be calculated according to the height of the cameras and their range of vision, arranging them in such a way that any point in the entire forest can be seen by three towers at the same time.

Fire control systems generally include firefighters with all the necessary combat equipment, including vehicles, machinery and tools, complemented by militias trained in combat aid and support for combatants, including ambulances, first aid equipment and medicines for people suffering from burns and accidents in the field while fighting fires.

16 FOREST MANAGEMENT

Geographic information systems technologies are increasingly being used to assess and monitor forest resources (PERSSON AND JANZ, 2010). The mapping, monitoring and management of forest spaces is carried out using technologies associated with terrestrial mapping and aided by satellites and *drones*. The English word *drone* means buzzing.

Terrestrial mapping still remains the most accurate, but technology with unmanned aerial vehicles (UAVs) is advancing and it can be predicted that topographic surveys will fall into disuse, just like the analog theodolites of the past. A drone is an unmanned aircraft, a flying robot, which can be controlled remotely or can fly autonomously by means of flight plans controlled by software in embedded systems, working in conjunction with on-board sensors and GPS.

Satellite images have a resolution of up to 0.5 meters per pixel and drones can provide images with millimeter resolution. Satellite sensors are more sophisticated, allowing them to assess a multitude of occurrences both on the ground and in vegetation.

The drones' high-resolution images make it possible to carry out forest inventories, identify areas affected by diseases, pests and environmental disasters, patches of soil with various problems, erosion, monitor and survey forestry operations, identify environmental problems, identify degraded areas and monitor environmental recovery areas, among other things.

16.1 Geographic Information System and Image Processing

A Geographic Information System is a computer-assisted system for the digital storage of maps and tabular data associated with mapping resources, which make it easier to produce customized maps, perform specialized database queries, analyze complex relationships, apply models and assist in decision-making. Geographic Information Systems for forest management can be classified into two categories:

1. resource inventory and monitoring;

2. analysis, modeling and forecasting for decision-making.

The development of a GIS for forest management should incorporate activities at two different stages of development (CRAIN and MACDONALD 1983).

Spatial data entry, editing and simple maps characterize the inventory and monitoring stage. In the modeling phase, overlays, reclassifications and suitability analyses are part of the decision-making process. The boundaries between these categories of activities are not distinct. Monitoring also includes analytical components to assess change or the outcome of specific interventions (McKENDRY et al, 1995).

The main data sources that feed a forestry GIS are:

- Maps (digitized manually or using *scanners*);
- Remote sensing products (aerial photographs, satellite images, videography);
- Forest inventories;
- Forest registers;
- Soil survey;
- Costs, simulation and linear programming; and
- GPS (Global Positioning System).

An image processing system is designed for the computer-assisted enhancement and interpretation of remote sensing images. With the advent of satellite systems with high-resolution images oriented towards natural resources, image processing techniques have been developed to convert images into classified maps. The analytical procedures of Image Processing Systems are consistent, accurate and fast, significantly aiding the development of thematic maps such as vegetation cover and soils.

Many software packages offer both features: GIS and Image Processing, which are complementary technologies and are now treated together as parts of a single system.

Resource assessment activities include: 1) inventorying the forest resources available for harvesting, fuel, food, recreation or conservation purposes, together with related data such as topography, soils, roads and hydrology, 2) monitoring changes that occur in these resources over time and 3) evaluating the potential productivity of the land for forest types, taking into account certain biophysical and climatic factors. It is in the assessment of forest resources that other technologies related to GIS systems, remote sensing and global positioning make direct and substantial contributions (McKENDRY et al, 1995).

McKendry et al (1995) list the following as the main joint applications of the two technologies in forestry:

- Resource Evaluation

– Inventory;
– Monitoring;
– Deforestation;
– Damage caused by pollution and pests;
– Assessment of suitability and productivity.

- Resource management

– Wood harvesting;
– Wood supplies for fuel;
– Fire management;
– Management of various resources.

According to McKendry et al (1995), the role of GIS in forestry can be understood as

> *1. GIS applications can benefit greatly from remote sensing and image processing technologies. Forests are complex assemblages of species that lend themselves well to broad-level inventory by means of remote sensing. However, the need for strong ground truth remains paramount and satellite positioning systems (such as GPS) are likely to play an important role in augmenting traditional forest survey activities.*

> *2. Forests are a dynamic resource, affected by many competing ecological processes and direct management interventions. Simulation modeling has been applied in forestry to a degree that is substantially higher than in many other disciplines. Simulation or process modeling is one of the most challenging areas of GIS applications and this activity is likely to increase as research and tools to support this type of application become more prevalent.*
>
> *3. It is clear that, throughout the world, forests are subject to many demands. As a result, many forest management problems are in the nature of multi-objective planning procedures. Unfortunately, GIS is not well developed for multi-objective planning. Stronger tools are needed for the analytical resolution of conflicting suitability and resource allocation choices.*
>
> *In a sense, forestry applications embody the full scope of GIS technology. Thus, their study provides an excellent overview of the state of the technology and its potential as a management tool for natural resource concerns (McKENDRY et al, 1995).*

16.2 Use of LiDAR technology in forestry

LiDAR (Light Detection and Ranging) is a remote laser scanner with its own energy source, which emits laser beams in the near-infrared band associated with global positioning technology, capable of modeling surfaces in three-dimensional form and precise positioning. The "Ranging" of LiDAR means the distance of something to be located from some operating point, i.e. LiDAR is the detection of objects by means of a remote laser beam, associated with global positioning systems (GPS).

Some of its first applications were in forestry. The laser beam pulsed by LiDAR penetrates the forest like sunlight, reaching different levels of penetration with each pulse until it reaches the ground, creating 3d images of the surface and what's above it. LiDAR makes it possible to estimate dendrometric and forestry variables, such as: diameter, height, crown dimensions, volume, aerial biomass and forest cover thickness.

The technology makes it possible to create digital models of the terrain without cover and of cover such as forests and trees. It can also be used at ground level to model objects such as trees. It has been used to carry out three-dimensional topographic surveys and to characterize vegetation structure quickly and with high precision. Leading companies in Brazil are starting to use LiDAR to replace many tasks that used to be carried out directly in the field, such as mapping, monitoring the formation of new plantations and reforming stands, detecting diseases and pests and even in forest inventories, monitoring the evolution of plantations.

17 FORESTRY PLANNING

Planning is the prior organization of future activities based on past knowledge in order to achieve objectives.

When drawing up a plan, you need to know the problem perfectly, define objectives, develop a plan of actions to achieve them, execute the plan, monitor its development and act on the results, reinforcing what is positive and correcting deviations and errors.

Planning is closely linked to Dr. Shewhart's PDCA. The PDCA cycle, which stands for *Plan, Do, Check, Act,* or "Plan, *Do, Check* and Act", is Dr. Shewhart's cycle, known in Japan as Dr. Deming's cycle, which spread in the country (MIRSHAWKA , 1990).

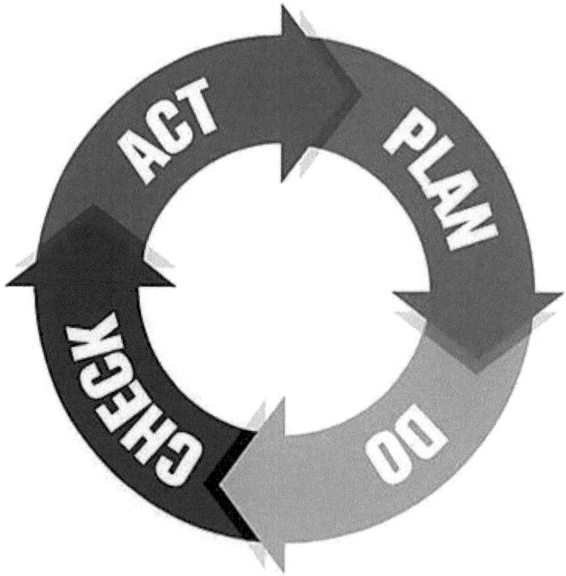

Figura 62 - Dr. Shewhart's PDCA.

Nowadays, planning is aimed not only at organizing activities, but also at continuous improvement and this is where the PDCA cycle comes in handy, as it involves reviewing and making decisions that lead to improved activities.

Planning is an activity carried out at four organizational levels related to organizational efficiency, effectiveness and excellence, which are achieved through the continuous improvement of the four organizational levels: resources, processes, relationships and identity.

Efficiency is achieved when resources and processes are optimal. Effectiveness is achieved when, in addition to optimal resources and processes, relationships are also optimal. Excellence is only achieved when, in addition to optimal resources, processes and relationships, the organizational identity is also up to the best standards.

17.1 Organizational levels

- Resources - Resources are related to the quality and quantity available for production. They are the contingent of personnel, property, facilities, equipment, stocks and capital;
- Processes - These are the flows that exist in an organization, be they material, information, personnel, transport, production, sales or money; they are closely related to the knowledge and technological level acquired and used in an organization; together with resources, they make up the operational planning level;
- Relations - This is the tactical planning level and concerns the type of leadership adopted by the organization, worker safety and health, the form of

internal and external communication; it is executive management and human relations planning;

- Identity - This is represented by the organization's history and its values and criteria, its philosophy and character, and is characterized by the type and quality of the products and services it provides, the breadth of its market, its environmental management and the quality of its internal and external relationships; it is the level of strategic planning where the organization's purposes, objectives and general policies are determined.

The organizational levels relate to the employees' personal knowledge and skills needs, to the responsibilities at each level and to the planning level as shown in Figura 63 .

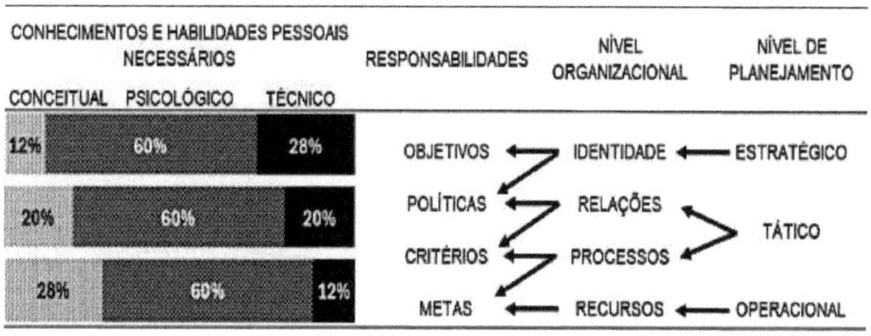

Figura 63 - Personal relationships, responsibility and organizational level pertinent to the planning level.

17.2 Planning phases

The main phases of planning are as follows:

1) Identify the object of planning;
2) Create a vision of the subject;
3) Define the planning objective;
4) Determine a mission or commitment to achieve the

goal;

5) Define work policies and criteria;
6) Set goals;
7) Develop an action plan to achieve the goals and fulfill the mission and objectives;
8) Establish a system for monitoring, controlling and analyzing planned actions;
9) Define an evaluation system for controlled data;
10) Provide for measures to be taken to prevent and correct any deviations and errors that may occur during the execution of the plan.

17.3 Characterization of planning

Planning is characterized as follows:

1) Dimensions;
2) Classes or types;
3) Phases or stages.

Whenever you talk about planning, more than one person is involved. Even if it's a job done by just one person, there will always be someone to provide them with something and the result of their work will serve someone else. So, in planning, you always think in terms of organizing the work of a team and that implies someone coordinating people who carry out tasks to achieve a goal.

17.4 Dimensions of planning

The dimensions of planning are:

- The subject or object of it;
- The elements involved;

- The organizational unit;
- Its implicit and explicit characteristics.

Subject or Object

It is the central theme of planning, for example:

- Environmental management system;
- Preservation and conservation areas;
- Forestry improvement;
- Wood production;
- Wood transportation;
- Transportation routes;
- Competition weed control;
- Pest and disease control;
- Seedling production.

Elements

The following nine planning elements are usually explicit in the plan:

- Purpose - What to do;
- Objectives - Why do it;
- Deadlines - How soon;
- Policies - What rules to follow;
- Criteria - How to judge;
- Procedures - How to do it, what steps to follow (action plan);
- Resources (technological and financial) - What to use;
- Monitoring - What to measure;
- Control - How to analyze and review what you have done.

ORGANIZATIONAL UNIT

The types of organizational units can be

- Corporate - Involves an entire institution;
- Subsidiary - an organization subordinate to another, also called a controlled company;
- Division - area, sector or department;
- Group or Commission - Involves the work of a team of people to achieve a goal;
- Operation - Involves the production of a single product or the performance of a single service;
- Project - The production of something, with its own objectives, timetable and budget for its execution, acting as an independent and formally constituted unit.

FEATURES

Characteristics of a plan can be implicit or explicit:

- Implicit characteristics - these are inherent to the subject of planning, they are those that are not declared but are automatically part of the plan;
- Explicit characteristics - are those that are stated in the plan as: a) Detail - strategic, tactical or operational; b) Quality related to the objective/object; c) Quantity to be achieved - related to the resources and processes required.

17.5 Class or type of planning

A plan can be classified according to

- Object of planning;
- Level of detail;
- Time limits or duration;
- Territory or catchment area;
- Number of criteria involved.

17.5.1 Level of Detail

It refers to the organizational level that is being planned:

- Strategic - general objectives and policies, organizational mission and vision;
- Tactical - is carried out by areas of activity or by departments at management level;
- Operational - is carried out by sector or product, detailing the amount of resources needed to achieve the objectives.

17.5.2 Deadlines

The timeframe refers to the planning horizon (period for which planning is carried out), and can be:

- Short - annual, usually at operational level;
- Medium - 5 to 10 years, usually at a tactical level;
- Long - more than 10 years, usually at a strategic level.

17.5.3 Territory or Space

Refers to the spatial scope of the plan, such as:

- Global - long/medium-term strategic;
- Continental - long/medium-term strategy;
- Bloc of countries - long/medium-term strategy;
- National - long/medium term strategic;
- State - Medium/short-term strategic/tactical;
- Municipal, or local - Medium/short-term strategic and/or tactical;
- Urban Area, Rural Area, Conservation Unit, Rural Property, Company - Strategic and/or tactical and/or medium and/or short-term operational.

17.5.4 Number of criteria

Criteria are pre-established limits for characteristics and targets:

- Monocriteria or linear objectives;
- Multicriteria or parallel objectives.

Multi-criteria planning deals with the simultaneous optimization of more than one alternative to achieve an objective, or the simultaneous optimization of more than one objective.

Forestry planning is characterized by multi-criteria.

17.6 Planning phases

When developing a plan, you first identify the subject or object of planning, then you need to gather all the information on the subject and predict what the four stages of the Deming cycle (PDCA) will look like; you also need to follow the sequence of the decision-making process, according to the situation analysis, which is mandatory and is the first stage, and the others will be carried out, or not, depending on this.

The phases of the decision-making process are: Situation Analysis (SA); Problem Analysis (PA); Decision Analysis (DA); and Potential Problem Analysis (PPA). After this procedure, you can move on to drawing up the plan, which is the document resulting from the planning. These phases each have their own development methodology and relate to the PDCA stages as shown in Tabela 80 below.

Merging the two methodologies: PDCA and Decision Process, we have the basic procedure for drawing up a plan.

Program the stages of the PDCA according to the phases of the decision-making process, which can be:

- Identify the subject or object;
- Gather information on the subject;
- Program the stages of the PDCA according to the phases of the decision-making process.

Tabela 80 - PDCA stages and phases of the decision-making process

	Stages	Phases of the decision-making process			
		Situation Analysis	**Problem Analysis**	**Decision Analysis**	**Analysis of Potential Problems**
	Identifying and gathering information	X	X		
P	1ª stage - Planning			X	X
D	2ª step - Execute			X	
C	3ª step - Evaluate	X	X		
A	4ª Step - Correct			X	X

The planning phases are as follows:

- Identifying and gathering information;
- Situation Analysis;
- Problem Analysis;
- Decision Analysis;
- Analysis of Potential Problems;

- Drawing up the Plan;
- Control and Monitoring System.

17.6.1 Identifying and gathering information

At this stage it is necessary:

- Establish the theme or subject - in this case it relates to the purpose or destination of the forest area involved - Example: lumber production;
- Identify the object of the planning - the forest or forests in question;
- Gathering information - everything about the environment (social, economic and natural) and the forest itself, which is the object of the planning.

17.6.2 Situation analysis

Decision analysis is made up of the following activities:

- Recognizing situations;
- Breakdown of situations;
- Determining the analysis sequence.

Situation recognition

This is the phase in which you define whether the situation requires planning, by answering:

- Is it a mandatory legal issue?
- Management on the subject requires?
- Do the organization's guidelines require it?
- Are there no alternatives for planning?
- Is it so important?
- Will there be big consequences if you don't plan?
- After all... is planning really necessary?

If the answer to the last question is yes, planning should be carried out.

The situation analysis is carried out according to the situation analysis flow (Figura 64).

Usually, forestry planning is not a problem in itself, but an issue. However, any deviation for the worse from what is desired, whether due to changes or comparisons with similar organizations, becomes a problem.

For example:

- Production costs that are too high in relation to the market can be considered a problem.
- The causes may not be known;
- Then you have to go through the phases of problem analysis, decision analysis and potential problem analysis.

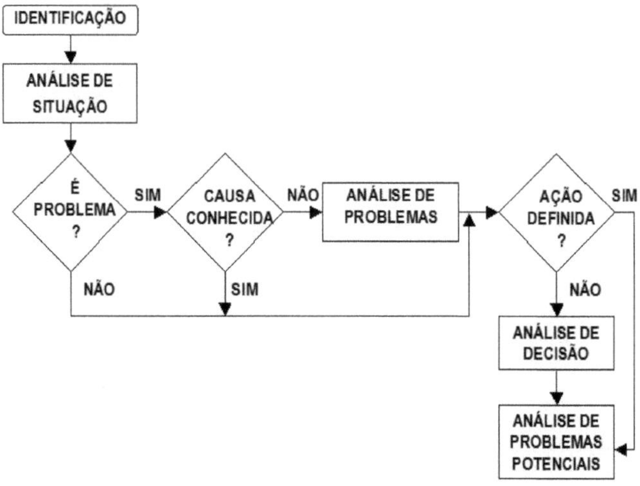

Figura 64 - Situation analysis flow.

17.6.3 PROBLEM ANALYSIS

The stages of this phase are as follows:

- Draw up the problem statement;
- Specify the problem;
- Check what deviations/changes have occurred;
- Identify the potential causes of deviations;
- Determine the most likely causes.

PROBLEM STATEMENT

It consists of the object and the problem itself (defect or deviation that the object presents).

PROBLEM SPECIFICATION

Start the specification by answering: what, where, when and how much is happening? - Then follow the script in Tabela 81 for the detailed specification of the problem.

Tabela 81 - Roadmap for problem specification

It's a problem It's bad We don't want	Not a problem Okay We wish
Which object is a problem?	Which object has no problems?
What problem does the object present?	What problem doesn't the object present?
Where does the object have a problem?	Where does the object not present a problem?
Where is the problem in the object?	Where is the problem not in the object?
When does the object have a problem?	When the object has no problem?
When does the problem present itself in the object?	When the problem isn't with the object?
How many objects have a problem?	How many objects don't have a problem?
How many problems does the object present?	How many problems does the object present?

CHECKING CHANGES

This is done by cross-referencing the problem specification questionnaire with checklists, process flowcharts, etc.

IDENTIFICATION OF POTENTIAL CAUSES

A cause and effect diagram, when compared with the changes that have taken place, makes it easier to determine when there was a change - it is the cause indicator.

- Testing the causes - Checking the causes can be:
- Factual - through available data on previous events;
- Real - through laboratory tests or examinations;
- Results - obtained through the implementation of mitigating or corrective measures with subsequent evaluation of the results.

We proceed in the following sequence:

- Define the wording of the decision;
- Establish decision guidelines;
- Find and list decision alternatives;
- Determine weights and limits for the guidelines;
- Compare the alternatives to the guidelines;
- Assess the risks of the alternatives chosen;
- Determine measures to minimize or compensate for risks.

The statement of the decision is formulated by declaring an action regarding an object and one or more limitations regarding the decision. The purpose of the decision or central theme is our object, it is the focus of the decision. The action stimulates the object of the decision; the limitations restrict the decision and identify it; the components of the statement must allow a group of comparable alternatives to be produced through mandatory or desirable guidelines.

Alternatives are the options available for decision-making and must be based on:

- Statement of decision;
- Breakdown of the statement;
- Guidelines;
- Adverse consequences of the decision.

Alternatives are compared on the basis of established guidelines. The guidelines or criteria are the restrictions we place on the choice of alternatives. They are the prerequisites used when choosing alternatives for the decision to be made. They can be mandatory (eliminatory) or desirable (qualifying) and can be evaluated by criteria:

- Qualitative - generally eliminatory;
- Quantitative - generally classificatory.

Mandatory guidelines are eliminatory and indispensable. They are attributes whose presence or absence, or whose type, determines whether or not an alternative is eliminated. For parametric guidelines, upper and/or lower limits are assigned from which the alternative becomes desirable or undesirable in relation to the guideline (Tabela 65).

The desirable guidelines are classifiable, negotiable and of relative importance, receiving different weights in terms of their relative importance. The weights for each of the desirable guidelines must be multiplied by a score assigned to each alternative in relation to meeting each guideline. The weighted average of each alternative is then calculated in relation to the weights in order to compare them (Tabela 66).

Once the alternatives have been ranked, the potential risks of each one are assessed. Potential risks refer to the likelihood of something wrong happening as a result of the decision taken and the seriousness of the situation, on a scale of 0% to 100%, for example. The analysis of potential problems helps to identify the risks of decisions. Planning is usually only carried out for the decision with the lowest risk.

17.6.4 Analysis of potential problems

The analysis of Potential Problems deals with the identification of future problems and is carried out in the following steps:

- Identify the process;
- Identify potential problems;
- Assess the risks of potential problems;
- Identify the causes of potential problems;
- Determine mitigating, protective and preventive measures;
- Drawing up the plan.

Process identification

Try to define the process by answering the following questions:

- What do we have to do or accomplish? (What is the process);
- Where? (What environment and facilities are involved in the process);
- When? (What deadlines, dates and times are involved);
- How much? (What are the goals of the process);
- How? (How the process is carried out in each phase).

Next, it's necessary:

- Develop a process flowchart (Figura 65);
- Develop a cause and effect diagram (Figura 66).

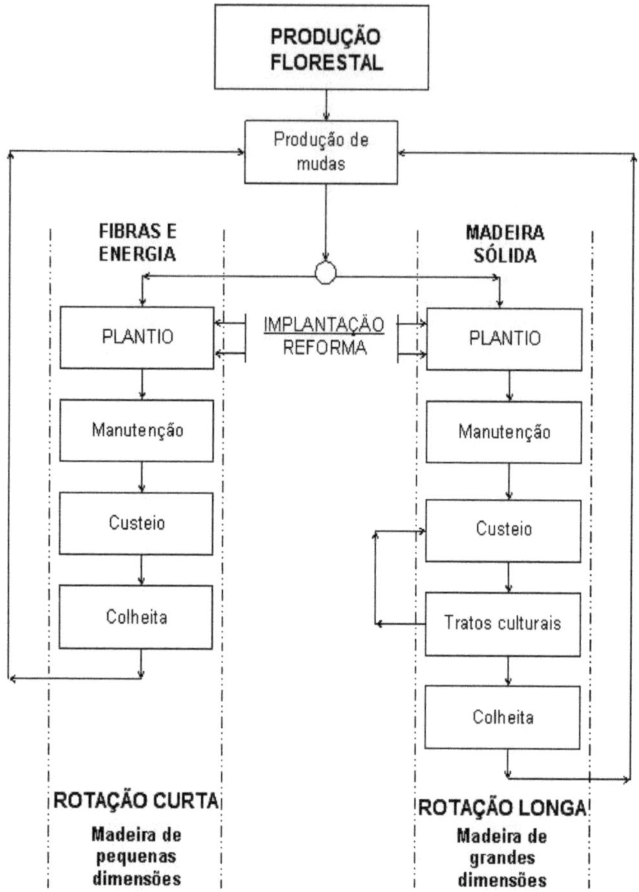

Figura 65 - Example of a forestry production process flowchart.

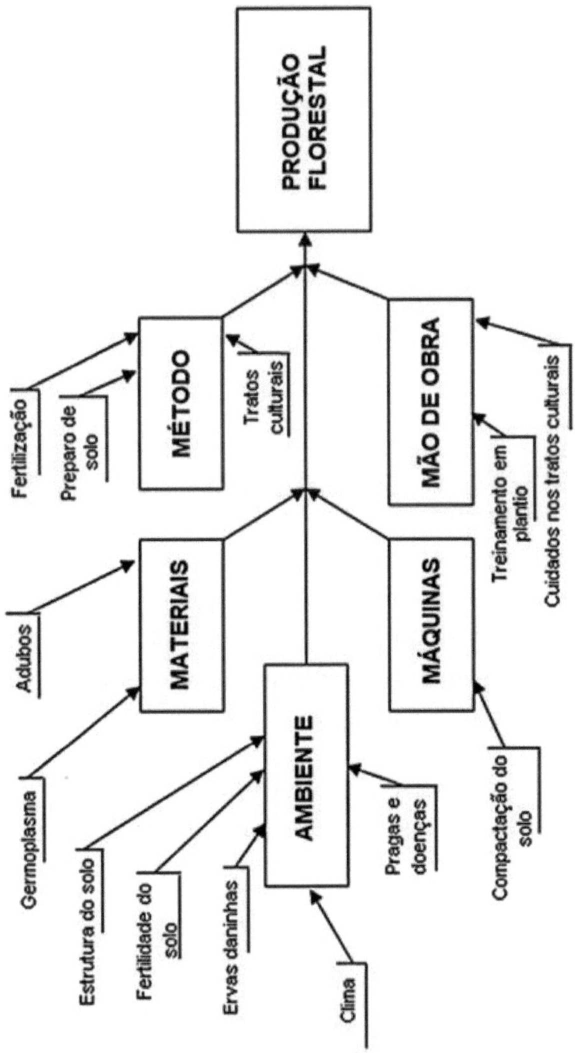

Figura 66 - Example of a cause and effect diagram for forestry production

IDENTIFYING POTENTIAL RISKS

Process flowcharts are used to identify potential risks by cross-referencing them with cause and effect diagrams.

Cross-referencing the process flowchart with the cause and effect diagram:

- It helps to identify potential problems and their causes;
- Facilitates risk assessment;
- It helps to identify measures to be taken to reduce risks.

By cross-referencing the flow with the cause and effect diagram of forestry production, for example, risks can be predicted from:

- Materials - Low quality germplasm; low quality fertilizer;
- Methods - Subsoiling at insufficient depth; planting without contour lines; etc;
- Labor - Knowledge, training, care;
- Machinery - soil compaction, subsoiling depth;
- Environment - fertility, competition, climate.

RISK ASSESSMENT OF POTENTIAL PROBLEMS

The risks in relation to potential problems are only 3:

- I can't do it;
- Doing it wrong;
- It takes too long to do.

Score each of the above risks in relation to the probability and severity of each potential problem identified and calculate the average = risk index.

QUALIFYING THE RISKS

Potential risks are assessed in terms of probability and severity (Tabela 82), such as:

- **Probability** - is the chance of happening defined as:
    - (A) High - it's going to happen;
    - (M) Average - can happen;
    - (B) Low - unlikely to happen.
- **Severity** - is the damage caused if it happens:
    - (A) High - very important;
    - (M) Average - important;
    - (B) Low - negligible.

Tabela 82 - Example of a risk assessment table

Risks	**Probability**			**Gravity**		
(adverse consequences)	**A**	**M**	**B**	**A**	**M**	**B**
First adverse consequence						
Second adverse consequence						
...						

Where: A = high; M = medium; B = low. Mark with an X the correspondence for the likelihood and severity of adverse consequences and score as shown in Tabela 83 to find the risk level.

Tabela 83 - Risk level (1 to 10)

Degree of comfort	Risk level	Probability	Gravity
I have no comfort	10	High	High
I have no comfort	7	High	Average
I'm comfortable	4	High	Low
I have no comfort	8	Average	High
Watch out	5	Average	Average
I'm comfortable	3	Average	Low
Watch out	6	Low	High
I'm comfortable	3	Low	Average
I'm comfortable	1	Low	Low

Actions on risks

Once the best alternative with the lowest risk index has been chosen, the planning of actions on the risks begins, seeking to eliminate the discomfort over the decisions. These actions can be

- Protective actions - actions on the effect to reduce the severity if it happens
- Prevention actions - actions on the causes to reduce the likelihood of this happening.

17.6.5 Process monitoring

By characterizing the process and identifying the risks, it is possible to determine what to control and how to analyze the measurement results. All the activities in the process, especially those involving the greatest risks, must be controlled using the appropriate means for each one.

17.7 Drawing up the Forest Management Plan

This is the last phase of planning. To start drawing up a plan, you need to:

- The situation is a problem and deserves attention;
- The problem is cyclical;
- The causes of the problem have been identified;
- The solutions to the problem have been defined.

It should be borne in mind that any plan has an expiration date and it is necessary to foresee when it will have to be revised.

Production planning traditionally resulted in the management plan. After the creation of forest management certification programs, the management plan became just a chapter in the Management Plan.

The management plan is made up of the timber production plan, which forecasts the production capacity of the management units over a planning horizon.

The management plan is the part of the Management Plan that is developed exclusively by Forest Engineers.

17.7.1 Forest Management

It is the planning of activities in the forest area over a period of time, to ensure that forestry operations are ongoing and within established parameters in order to achieve the physical objectives and goals of sustainable production.

Sorting is classified as:

- Rural properties;
- of production activities.

Spatial planning

It is the process of organizing space according to its capacity and vocation in a sustainable way, including:

- Areas of ecological, paleontological, archaeological, etc. interest;
- Legal reserve areas;
- Permanent preservation areas;
- Infrastructure areas and access roads;
- Productive areas.

The distribution of production units and subdivisions within the space is generally carried out in hierarchical units as follows:

- District: several nearby areas, without continuity, under the same administration - 5 to 20 thousand ha;
- Farm/Horto: contiguous areas within a district, made up of several production units, including non-productive areas - 50 to 5 thousand ha;
- Production units - are made up of plots with a standard size associated with accounting cost centers;
- Plot: production unit in a contiguous area, with well-defined natural and/or artificial boundaries - 5 to 50 ha;
- Section: production unit with similarity of site and homogeneous productivity - subdivision of the plot - area 0.5 to 5 ha.

The objectives of spatial planning are mainly to facilitate planning at each unit level for forestry activities such as:

- Silvicultural treatments: thinning, fertilization, choice of species, improvements;
- Safety: separating stands reduces the risk of damage, whether biotic (insects, fungi) or abiotic (fire), and facilitates mobility;
- Logistics: roads and accessibility, silvicultural execution, exploitation, renovation (natural or artificial).

Spatial planning makes the forest more visible and controllable through maps and registration of the units.

Spatial planning in new areas should be carried out before planting, in the following order:

1) Preparation of the planialtimetric map of the area;
2) Demarcation of areas restricted by legislation;
3) Demarcation of mandatory crossing points for transport routes and location of bridges and culverts, seeking to minimize the environmental impact of construction;
4) Location of borrowing areas for materials to be used in road construction;
5) Surveying and mapping the sites;
6) Layout of the plot network, with: roads, size, distance between them;
7) Limiting the sections and layout of the extraction routes.

In spatial ordering, it is recommended that all levels be in Arabic numerals to facilitate computer programming, which can be as follows:

- Section - unit digits (0 to 9);

- Slot - digits of the tens and hundreds (10 to 990);
- Production Unit - thousand to hundred thousand digits (1,000 to 999,000) - is the work unit that will have an accounting cost center;
- District - digits from a million to a hundred million (1,000,000 to 999,000,000);
- Region - billion digit (1,000,000,000).

Example: 3.021.342.075 corresponds to section 5, plot 7, production unit 342, district 21, region 3.

TIMING

Temporary forest management is a plan that defines the process of operational activities for forest production, in order to meet the established goals, taking into account:

- Seedling production;
- Renovation and new plantings;
- Forest maintenance;
- Cultivation - pruning and cleaning cuts;
- Timber harvesting - thinning and final felling;
- Forestry transportation.

The characterization of stands for silvicultural planning must take into account and identify the following characteristics of the stands:

- Species;
- Age: years old or since renewal. Equine or Inequine;
- Verticality: regular (strata) or irregular;
- Horizontality: spatial occupation by trees (sparse, clear, open, closed, very closed);

- Homogeneity - represents the percentage of participation of each species: a) in artificial stands (monoculture, groups, strips, rows, groves); b) in natural stands (random, tendency to cluster, clustered).

The sequence for planning the production of planted forests can be carried out in the following stages:

1) Determining the possible rotation ages for the forest;
2) Model dominant height growth and Site Index (SI) tables using Assmann's dominant height [$h_{100} = f(t)$];
3) Model the growth of the average height [$h = f(h_{100})$];
4) Model the growth of the average diameter [$d = f(h100)$];
5) Model the distribution of diameters over time;
6) Model the variation in tree shape [shape = f (d, h, t)];
7) Determine the variation in stand density according to tree size, site and age;
8) Establish the thinning regime, if any;
9) Assess the best age for rotation;
10) Determine the productivity and production of stands (develop production tables);
11) Sorting production - sorting the cuts according to the production target.

Rotation is the period of time from the renewal of the stand to the final cut. It is defined according to the type of product offered and economic, social and environmental objectives.

There are various ways of determining the rotation of forest stands that define different types, such as:

- Technical rotation - Corresponds to the age at which the tangent to the production curve is maximum; it is the age at which ICA and IMA are equal;
- Rotation of maximum net forest yield - this is the age at which the greatest profit occurs, i.e. when the difference between income and cost is maximum (highest EAV);
- Rotation of maximum net yield of the land - this is the age at which the highest VET occurs;
- Socio-economic rotation - this is the age at which the greatest value is provided in relation to the multiple use of the forest.

17.8 Forest Management Planning

Certification systems have established principles, criteria and indicators for organizations to obtain certification. The principles are the social, economic and environmental problems related to the forestry production process. Criteria are developments of the principles, establishing maximum and minimum limits for their evaluation. Indicators are the important parts of each criterion that must be monitored; they can be quantitative or qualitative.

17.8.1 Principles and criteria

Forest certification programs establish principles similar to those listed below:

- Compliance with legislation and standards;
- Property rights;
- Respect for local communities;
- Respect for employees;
- Respect for the environment;

- Economically, environmentally and socially sustainable production;
- Monitoring and Evaluation;
- Social, environmental and economic support for the region in which it operates.

The sequence for drawing up the plan could be as follows:

1) Identify the object of planning;
2) Define the planning objective;
3) Establish a mission and create a vision for the enterprise;
4) Define working principles and criteria;
5) Set goals;
6) Develop an action plan to achieve the goals and fulfill the mission and objectives;
7) Establish a system for monitoring, controlling and analyzing planned actions;
8) Define an evaluation system for controlled data;
9) Economic and financial analysis of the project;
10) Provide for measures to be taken to prevent, mitigate and correct any deviations and errors that may occur during the execution of the plan.

17.8.2 Identifying the Planning Object

The object is Forest Management for the production of what is to be supplied or produced. Identifying the object of planning involves:

- Identify the organization;
- Define the product to be supplied;
- Locate the enterprise;

- Describe the socio-economic and biophysical environments;
- Characterize the production chain involved;
- Describe the market.

Identification and Scope

The identification is listed:

- Shareholders and quotas;
- The location must include the address and contact details of the organization's headquarters;
- The coverage area should be described in terms of timber production, infrastructure, conservation and preservation, industrial production and commercial coverage.

Socio-economic environment

The components involved in the socio-economic environment are as follows:

- Internal and external actors involved;
- Socio-economic characteristics;
- Development policies;
- Public infrastructure;
- Benchmarking.

Biophysical environment

The biophysical environment mainly involves:

- Physiography;
- Natural vegetation;
- Land use and cover;
- Conservation units;
- Environmental zoning;
- Soil types and fertility;
- Land availability and occupation trends;

- Hypsometry;
- Land slope;
- Hydrology;
- Climate.

PRODUCTION CHAIN

The main forestry production chains are:

- Solid wood industries;
- Pulp industries;
- Wood energy production;
- Charcoal production;
- Panels;
- Furniture;
- Resins and wood extractives.

17.8.3 OBJECTIVE

It's what you want to be or do. The main objective is usually related to the species and purpose of the wood. Example: "Production of pine wood for sawmills".

The objectives must be set:

- Wood production;
- Economic;
- Social;
- Environmental.

17.8.4 MISSION AND VISION

It is important to state the mission and vision for the future of the organization. This helps to establish working policies.

Mission is the organization's commitment to society, its reason for existing; it should aim to satisfy some need of society. Example: "To supply society with quality wood for construction and furniture in a sustainable way".

Vision is the organization's perspective on the subject addressed in the plan. Example: "To be an international benchmark for sustainability and quality in the production of sawn timber".

17.8.5 Principles

These are the guidelines or work policies. Most are pre-established in the forest certification standards and characterize the organization's approach:

- Relationships within the organization and with customers and suppliers;
- Your processes;
- Resources and technology;
- The environment;
- To society.

17.8.6 Goals

A goal is a quantified objective with a deadline. Goals refer to what the organization plans to do during a given period in relation to the objectives. Targets are established qualitatively and quantitatively with reference:

- To production;
- To the collaborators;
- To customers and suppliers;
- Production processes;

- The resources and technology used;
- The economy;
- The environment;
- To society.

Some examples of targets are given below:

- Production target - To produce 100,000 m^3 of wood for sawmills during the next financial year;
- Environmental target - Reduce the use of forestry inputs with polluting potential by 20% over the next financial year;
- Economic target - Reduce the cost of timber production by 5% over the next financial year.

Physical Production Targets are the physical results expected in relation to production and stands in order to achieve the objectives. They should be established for production, stock and stand renewal as:

- Production target - volumes of assortments to be produced;
- Stock level target - desired stock level in relation to normal;
- Renewal target - percentage of stands planted, reformed or regrowing in relation to what will be harvested during the period in question.

Economic targets are the expected financial results. They are evaluated by financial indicators such as:

- NPV - Net Present Value;
- ETV - Expected Land Value;
- IRR - Internal Rate of Return;
- EAV - Annual Equivalent Value;

- C/B - Cost/Benefit Ratio.

Economic goals are affected by the fluctuating market, monoculture, climate, species, cutting order, cultivation, etc.

Speidel (1967) listed three criteria that can be followed when setting goals:

- Forecasting consumption and demand - future market and stocks - is done by national (IBAMA, Ministries) and international (FAO) agencies. It serves to guide future markets;
- Supply and production forecast - quantity to be placed on the market based on expected consumption and current stocks;
- Trade forecasting - comparison of supply and demand - distance from consumer centers and influence on prices - helps to establish the company's geographical location.

17.8.7 Action plan

The action plan must be carried out for each of the targets separately:

- Production - Forest management plan;
- Personnel - Training, safety and health;
- Social - Support and improvement of living conditions;
- Environmental - Impact reduction and environmental recovery;
- Economic - Reducing costs and debt, increasing profitability.

The action plan for production involves:

- Improving the genetic base;
- Soil conservation;

- Forest protection;
- Maintenance of infrastructure and access roads;
- Renewal of stands and new plantings;
- Cultivation and maintenance of stands;
- Sorting the harvest;
- Forestry transportation.

Each action must make the following explicit:

- What to do? - Action itself;
- Why do it? - reason;
- Where to do it? - Location, installation and structure;
- How to do it? - Method;
- What to do? - Equipment and machinery;
- What to use? - Materials and supplies;
- When to do it? - Deadlines, timetable;
- How much to do? - Goals;
- Who does it? - Personnel and training;
- Expected results.

The action plan can be presented as follows Tabela 84

.

Tabela 84 - Action plan for production

Acti ons	Reas on	Installat ion	Meth od	Machin es	Materi als	Dead lines	Tar get	Staff
What to do?	Why do it?	Where to do it?	How to do it?	What to do?	What to wear?	When to do it?	How much to	Who does it?
Acti on 1								
Acti on 2								
...					...			
Acti on n								

17.8.8 Expected results

The expected results for each action, both qualitative and quantitative, should be predicted and then compared with the monitoring results.

17.8.9 Monitoring, control and analysis

Monitoring must be carried out for each of the activities/actions and the situation of the resources, goods and services produced, such as:

- Personnel - Health, safety, training, etc;
- Management units - inventory, growth, pests and diseases, weather losses, soil fertility and structure, etc;
- Operations - production and productivity, equipment maintenance, training, etc;
- Infrastructure - roads, shipyards, etc;

- Conservation units - monitoring fauna and flora;
- Environment - soil and water contamination, invasive species, etc.

Monitoring involves identifying the objectives of what you want to control and listing the parameters and evaluation attributes needed to meet the established objectives.

It must be established:

- Criteria for evaluating parameters and attributes;
- Measurement and data collection system;
- Sample analysis system;
- Statistical data analysis system;
- Model for presenting monitoring results;
- System for disseminating monitoring results.

Control of the actions carried out must be short enough to detect and analyze any deviations from the plan so that corrective action can be taken quickly.

Monitoring aims to ensure that the plan is being carried out within the established criteria and standards. This is done by defining what, why and how to measure in order to control the implementation of actions.

17.8.10 Evaluation system

The evaluation system must provide for periodic reports to be drawn up and analyzed by higher authorities so that decisions can be made. It should include the analysis of deviations, which can be done according to the problem analysis method.

The periodicity of the plan's evaluation should be in accordance with the criticality of what is analyzed, internal and external legislation and standards, and should cover all planned actions.

The assessment system for certification purposes includes internal and external audits:

- Internal audits - these are carried out annually by trained internal staff, or by a specially hired consultancy;
- Pre-audit - usually carried out by a consultancy hired by the organization undergoing the certification process to filter out non-conformities in the system developed;
- External audits - these can be certification or maintenance audits, as follows:
    - Certification audits are carried out once the management system has been developed and the management plan has been drawn up;
    - Maintenance audits take place periodically according to the certifier's schedule, in order to maintain certification.

17.8.11 Economic and financial analysis

The economic and financial analysis must take into account

- Forest management units must have an independent accounting cost center;
- Cost and revenue analyses should be carried out no more frequently than quarterly;
- The economic and financial indicators should be analyzed annually (IRR, VAE, etc.).

17.8.12 Impact prognosis

Impacts must be foreseen:

- Economic;
- Social;
- Environmental.

17.8.13 Risk analysis

The risk analysis must be carried out for each action independently and the analysis model can be based on the potential problem analysis model.

17.8.14 Prevention and correction

Taking action on the risks should be considered in order to eliminate discomfort over the decisions made.

You can decide to live with the risks if they are very remote and of low severity, but in other cases you can decide to minimize or even eliminate them.

Types of actions to minimize or eliminate risks:

- Protective Actions - These are taken against the effects of risks in order to reduce their severity if they occur;
- Preventive Actions - These are taken on the causes in order to reduce the likelihood of this happening;
- Corrective Actions - These are actions that are pre-designed in case, or designed after, a deviation from what was programmed occurs.

17.8.15 Prognosis of results

The results must be predicted for all the areas and scope of the plan, such as:

- Social;
- Environmental;

- Economic;
- Collaborators;
- Customers and suppliers.

17.8.16 Plan Review

Management plans for Forest Certification purposes are reviewed in accordance with the technical standards of the certification systems. Certificates are usually valid for 4 to 5 years and, depending on the criticality of the system, the period can be shorter. The certifier may require the Management Plan to be revised at any time as a result of maintenance audits. The action plans that form part of the management plan are usually developed each year for budgeting purposes.

18 REFERENCES

ABNT. Interpretation NBR ISO 14001 (1996). Rio de Janeiro: ABNT-Cb-38/Sc- 01/Interpretation Group, July 2001.

ABNT. NBR 14789: Forest Management - Principles, criteria and indicators for forest plantations. Rio de Janeiro, 2001.

ABNT. NBR ISO 14001: Environmental Management Systems - Specification and Guidelines for Use:. Rio de Janeiro, 1996.

ABNT. NBR ISO 14004: Environmental management systems - General guidelines on principles, systems and support techniques. Rio de Janeiro, 1996a.

ABNT. NBR ISO 14010: Guidelines for environmental auditing - General principles. Rio de Janeiro, 1996b.

ABREU, J. C. et al. Structure and spatial distribution of andirobe trees (*Carapa spp.*) in floodplain forest of the Amazon estuary. **Ciência Florestal**, Santa Maria, v. 24, n. 4, p. 1009-1019, dezembro/ 2014.

AMARAL, W. A. N. **Recursos Florestais em Propriedades Agrícolas**: Seleção de Espécies/Procedências e Clones. [Piracicaba]: ESALQ/USP, 2016.

ASSMANN, E. **The principle of forest yield study** [Oxford]: Pergamon press, 1970. 506 p.

BAKER, F. S. **Principles of silviculture**. New York: McGraw Hill, 1950.

BARROS, R. S. M. **Measures of biological diversity**. Juiz de Fora UFJF/PGECOL, 2007. 13 p.

BERTOLOTI, G.; SIMÕES, J. W.; NICOLIELO, N.; GARNICA, J. B. Effects of different thinning methods and intensities on the productivity of *Pinus caribaea* var. *hondurensis* Barr. et Golf. **IPEF**, n.24, p.47-54, Aug-1983.

BHAMIDIPATI, N. L. **Rényi Entropy**. [s.l.]: MathWorld-- A Wolfram Web Resource. Available at: <http://mathworld.wolfram.com/ RenyiEntropy. html>. Accessed in 2009.

BRAZIL. **Law No. 12.651**, of May 25, 2012. Provides for the protection of native vegetation; amends Laws Nos. 6.938, of August 31, 1981, 9.393, of December 19, 1996, and 11.428, of December 22, 2006; repeals Laws Nos. 4.771, of September 15, 1965, and 7.754, of April 14, 1989, and Provisional Measure No. 2.166-67, of August 24, 2001; and makes other provisions.

BRAZIL. **Law No. 9.985**, of July 18, 2000. Regulates art. 225, § 1, items I, II, III and VII of the Federal Constitution, institutes the National System of Nature Conservation Units and makes other provisions.

BRAUM-BLANQUET, J. **Fitosociologia**: bases para el estúdio de lãs comunidades vegetales. Madrid: H. Blume, 1979. 820 p.

BURGER, H. The crown structure of echinaceous conifer populations. Mitt, Switzerland: **Anst. f. d. forstl**..verse 21, p.5-58, 1939.

BUSSAB, W.O. Analysis of variance and regression. São Paulo: Atual, 1986. 147 p.

CARVALHO, A.M. Valorization of wood from the *Eucalyptus grandis* x *Eucalyptus urophylla* hybrid through the joint production of small-sized sawn timber, cellulose and firewood. Dissertation (Master's Degree in Forestry Engineering) - University of São Paulo, Luiz de Queiroz College of Agriculture, Piracicaba, 2000.

CARVALHO, D.L.; LIMA, A.V. Metodologias para Avaliação de Impactos Ambientais de Aproveitamentos Hidrelétricos. In: XVI ENCONTRO NACIONAL DOS GEÓGRAFOS, Porto Alegre. 2010.

CARVALHO, H.M. de. **The Methodology of the Decision-Making Process. Linhares**: Postgraduate course in Strategic Human Resources Management, FACCL, 1997a.

CARVALHO, L.A. Road planning. **Opiniões Magazine** - Division F, Ribeirão Preto, a.13, n.43, P.11-12, mar-mai/2016.

CARVALHO, P.H.R. **Espécies arbóreas brasileiras**. Brasília: Embrapa Informação tecnológica; Colombo: Embrapa Florestas, 2003. 1039 p.

CARVALHO, P.E.R. **Brazilian forest species**: silvicultural recommendations, potential and use of wood. Colombo: EMBRAPA-CNPF / Brasília: EMBRAPA-SPI, 1994. 640 p.

CHIARADIA, T. ISO 14001 - Environmental Management System. IN: SEMINÁRIO ALSTON, Itajubá, Alston Brasil, 2001.

COELHO, V.C.M.C.; HOSOKAWA, R.T. Evaluation of growth reaction to thinning in Pinus taeda L. **Agroambiente**, UFR, Boa Vista-RR, v.4, p.42-48, jan-jun/2010.

CONTRERAS, M.A.; AFFLECK, D.; CHUNG, W. Evaluating tree competition indices as predictors of basal area increment in western Montana forests. **Forest Ecology and Management**, v.262, n.11, p.1939-1949, 2011.

COPEL. Important characteristics for a good integration of trees with the urban environment. Curitiba, 23/04/2009. Available at: <http://www.copel.com/hpcopel/root/pagcopel2.nsf/

CRAIN, I.K.; MACDONALD, C.L.. From land inventory to land management: the evolution of an operational GIS. *IN*: AUTOMATED CARTOGRAPHY: INTERNATIONAL PERSPECTIVES ON ACHIEVEMENTS AND CHALLENGES, Vol. 1, B.S. Weller, ed. Ottawa: Steering Committee for the Sixth International Symposium on Automated Cartography, p.41-50, 1983.

CRUZ, A.V. Estradas vicinais: abordagem pedológica, geotécnica, geométrica e de serventia de dois trechos rodoviários não pavadas no campos da UFV-MG. Dissertation (master's degree) - Federal University of Viçosa - Viçosa, MG, 2005.

DANSEREAU, P. The plans of biogeography. **Revista Brasileira de Geografia**, Rio de Janeiro, IBGE, v.8, n.2, p.189-210, abr-jun/1946.

DOSSA, D. et al. Application with profitability analysis for cultivated forest and grain production systems. Colombo: Embrapa Florestas, Documentos 39, 2000.

EATON, R.A.; GERARD, S.; DATTILO, R.S. A Method for Rating unsurfaced Roads. **Transportation Research Record**, n.1106, p.34-43, August-1987.

EXPRESSIVE. Magazine website. Available at: <http//:www.revistaexpressiva. com.br/>. Accessed on: 31/07/2005.

FAO. **Global forest resources assessment 2010**: terms and definitions. Rome: Food and Agriculture Organization of the United Nations - Forestry Department, Forest Resources Assessment Programme, Working paper 144/E, 2010.

FARO, C. **Elementos de Engenharia Econômica**. 3 ed. São Paulo: Atlas, 1979. 328 p.

FERREIRA, J.P. **Análise da cadeia produtiva e estrutura de custos do setor brasileiro de produtos resinosos**. 2001. 92f. Dissertation (Master's Degree in Applied Economics) - Escola Superior de Agricultura "Luiz de Queiroz", Universidade de São Paulo, Piracicaba, 2001.

FERREIRA, M. **Choosing tree species to form forest clumps**. Piracicaba: IPEF. Documentos Florestais. n.7, p1-15, jan. 1990.

FERREIRA, M. Choice of Eucalyptus Species. **Circular Técnica IPEF**, v.47, p.1-30, 1979.

FLORIANO, E.P. Inventory and prognosis of forest production at Todesflor in Cachoeira do Sul, Rio Grande do Sul. Cachoeira do Sul: Ambiente Inteiro, 2006.

GARCÍA, O. **Notes on forest mensuration**. Prince George, BC, Canada: UNBC/IFM, 2004.

GARCIA, P. O.; LOBO-FARIA, P. C. **Metodologias para Levantamentos da Biodiversidade Brasileira**. Juiz de Fora: UFJF/PEGCOL, [2007]. 22p.

GENNIA NEWS. Drawings of trees. Available at: <http://genniartist.blogspot.com/2016/05/desenhos-de-arvores.html>. Accessed in: 2018.

GIBSON, M. D.; CLASON, T. R.; HILL, G. L.; GROZDITS, G. A. Influence of thinning and pruning on southern pine veneer quality. In: ELEVENTH BIENNIAL SOUTHERN SIVICULTURAL RESEARCH CONFERENCE, Department of Agriculture, Forest Service, Southern Research Station, Athens-GA, Nov. 14, 2001. p. 163-167.

GRIMALDI, R.; MANCUSO, J.H. **Total Quality**. Issue 6. São Paulo: SEBRAE-FOLHA, 17/04/1994.

GROBER, U. Von Kursachsen Nach Rio - Ein Lebensbild über den Erfinder der Nachhaltigkeit Hannß Carl Edler von Carlowitz und die Wegbeschreibung eines Konzeptes - aus der Silberstadt Freiberg. Available at: <http://www.forschungsheim.de/fachstelle/arb_carl.htm>. Accessed on: 05/11/2002.

HENRIQUES, R. P. B. et al. Field methods in ecology: research projects produced in the seventh course. BRASÍLIA: UNB, 1996.

HORTA, J.C.O.S. Rehabilitation and Upgrading Design of Earth and Gravel Roads in Tropical Developing Countries. **Transportation Research Record**, Washington , n.1291, p.215-223, 1991.

HUSCH, B.; MILLER, C. I.; BEERS, T. W. **Forest mensuration**. 3rd ed. New York: John Wiley and Sons, 1982. 401 p.

IAMAZON. Floresta para Sempre: a manual for timber production in the Amazon. Available at: <http://imazon.org.br/floresta-para-sempre-um-manual-para-a-producao-de-madeira-na-amazonia/>. Accessed on: 05/10/2018.

IBÁ. Report 2017. Brasília: Brazilian Tree Industry, 2017. Available at: <https://iba.org/images/shared/Biblioteca/IBA_RelatorioAnual2017.pdf>.

IBA. Institutional website of the Brazilian Tree Industry. Available at: <https://www.iba.org/>. Accessed in: 2018.

IBGE. Survey and classification of land cover and land use: Retrospective study of phytosociological characteristics and forest potential in the state of Pará. Rio de Janeiro, 2007. 117 p.

IBGE. PEVS - Production of Vegetable Extraction and Forestry - Tables 2023. Available at: <https://www.ibge.gov.br/estatisticas/economicas/agricultura-e-pecuaria/9105-producao-da-extracao-vegetal-e-da-silvicultura.html?=&t=resultados>. Accessed on: 2024.

IB-USP. Ecological succession. [São Paulo]: Institute of Biosciences, University of São Paulo. Available at: <http://www.ib.usp.br/ecologia/sucessao_ecologica_print.htm>. Accessed in: 2009.

IDEIA. Environmental Impact Assessment Course - examination and assessment of environmental aspects and their impacts. Porto Alegre: IDEIA/CREA, 2002.

IMANÃ, J. E.; SILVA, G. F. da; PINTO, J. R. R. Age and growth of trees. **Forestry technical communications**, Brasília, UFB, Department of Forestry Engineering, v.7, n.1, 2005. ISSN 1517-1922. 43p.

IMAÑA, J.E.; SILVA, G.F.; TICCHETTI, I. Dendrometric variables. **Forestry Technical Communications**, Brasília, UNB-Faculdade de Tecnologia-DEF, v.4, n.1, October 2002.

ISO. ISO 31.000 - Risk management: principles and guidelines. Genoa, 2009.

IU. Testing normality using SAS, STATA, and SPSS. Indiana: Indiana University - IU, UITS - Center for Statistical and Mathematical Computing. Available at: <http://www.indiana.edu/~statmath>. Accessed on: 15/May/2004.

JOHNSTON, D. R.; GRAYSON, A. J.; BRADLEY, R. T. **Forest planning**. Lisbon: Calouste Gulbenkian Foundation, 1977.

KRAJICEK, J.E.; BRINKMAN, K.A.; GINGRICH, S.F. Crown competition factor, a measure of density. **Forest Science**, v.7, n.1, p.35-42, 1961.

KRAMER, P. J.; KOZLOWSKI, T. **Fisiologia das árvores**. Lisbon: Calouste Gulbenkian Foundation, 1972. 745 p.

LAAR, A.; AKÇA, A. **Forest Mensuration**. Dordrecht, The Netherlands: Springer, 2007.

LIMA, I. L. P.; SCARLOT. **Good management practices for the sustainable extraction of Mangaba.Brasília**: Embrapa Recursos Genéticos e Biotecnologia, 2010.

LOETSCH, F.; ZÖHRER, F.; HALLER, K. E. **Forest inventory**, Vol. II. Reinbek: Forest Inventory Section, Federal Research Organization for Forestry and Forest Products, 1973. 469 p.

LOPES, F. S.; WIPIESKI, C. J. Termelétricas: uma nova oportunidade de negócios na indústria madeira.Curitiba: STCP, inf. n.5, p.16-18, 2001.

MACHADO, S. A.; MELLO, J. M. de; BARROS D.A. de. Comparison of methods for evaluating total wood volume per unit area for Paraná pine in southern Brazil. **Cerne**, v.6, n.2, . p. 055-066, 2004.

MAGURRAN, A. E. **Ecological Diversity and Its Measurements**. Princeton, NJ: Princeton University Press, 1988.

MALEKI, K.; KIVISTE, A.; KORJUS, H. Analysis of Individual Tree Competition Effect on Diameter Growth of Silver Birch in Estonia. **Forest Systems**, v.24, 2ed.; e023, 13p., August 2015.

MATTEUCCI, S.D.; COLMA, A. **Metodologia para el estudio de la vegetacion**. Washington: O.E.A., 1982, 182 p.

McKENDRY J.E., EASTMAN J.R., ST. MARTIN K. AND FULK M.A. ***Explorations in Geographical Information Systems Technology***, UNEP/UNITAR, Clark University, Worchester, 1995.

MIRSHAWKA, V. The implementation of Quality and productivity by the method of DR. Deming. São Paulo: McGraw-Hill, 1990.

MOHR, R. R. **Failure Modes and Effects Analysis**. 8th ed. St. Louis (USA): Jacobs Sverdrup, Feb. 2002.

MOORHEAD, D. J.; DANGERFIELD, C. W. **Intensive forest management 2**. Tifton: The University of Georgia, College of Agricultural and Environmental Sciences and Warnell School of Forest Resources, 1998.

MÜLLER, I. Trunk shape and wood assortments for *Eucalyptus grandis* Hill ex Maiden, managed in high stem, in the Southeast Region of the State of Rio Grande do Sul. 135 f. Thesis (Doctorate in Forest Engineering) - Federal University of Santa Maria - UFSM, Santa Maria, 2004.

NAJBERG, S.; PEREIRA, R. de O. **Jobs generated by a R$ 10 million increase in production**. Rio de Janeiro: Sinopse Econômica, n.133, BNDES, Mar./2004. Available at: < http://www.bndes.gov.br/conhecimento/publicacoes/catalogo/sinopse.asp>. Accessed on: 28/05/2005.

NEMEC, A. F. L. Analysis of repeated measures and time series: an introduction with forestry examples. **Biom. Inf. Handb**. 6. Res. Br., B.C. Min. For., Victoria, B.C. Work. Pap. 15/1996. 90 p.

NEVES, G. A.; MARTINS, C. A.; MIYASAVA, J.; MOURA, A. F. **Economic-financial analysis of the exploitation of Pinus resinifera in small rural modules**. Specialization monograph in Agribusiness FEA/FIA/USP, Sorocaba, 2001.

NEWMAN, D.H. **The optimal forest rotation: a discussion and annotated bibliography**. Asheville, NC: U.S.D.A.-FS, Southeastern Forest Experiment Station, Gen. Tech. Rep. SE-48,1988.

NEWTON, A. C. **Forest ecology and conservation**: a handbook of techniques. Oxford: Oxford University Press, 2007. 454 p.

OLIVEIRA, T. Risk and forest management: an opportunity for efficiency. IN: V ENCONTRO NACIONAL, I CONGRESSO INTERNACIONAL DE RISCOS, Coimbra - 29/5/2009.

ORMOND, J. G. P. Glossary of Terms Used in Agriculture, Forestry and Environmental Sciences, 3 ed. Rio de Janeiro: BNDES, 2006.

PAMPLONA; E. de O.; MONTEVECHI, José A. B. **Engenharia econômica I**. Itajubá: UNIFEI/IEM, 1999.

PAYANDEH, B. Comparison of method for assessing spatial distribution of trees. **Forest Science**, v. 16, n. 3, p. 312-317, 1970.

PERSSON, R.; JANZ, K.. Assessment and monitoring of forest and tree resources. **Forest resources assessment**, [Rome], FAO, p. 16-29, 2000.

PIELOU, E.C. Mathematical ecology. New York: Wiley, 1977. 385 p.MUELLER-DOMBOIS, D.; ELLEMBERG, H. **Aims and methods of vegetation Ecology**. New York: Wiley, 1975. 547p.

PRODAN, M. **Mensura forestal**. San José: Inter-American Institute for Cooperation on Agriculture (IICA), 1997. 562 p.

REZENDE, J. L. P. **Course in the economic analysis of forestry projects**. Santa Maria: PPGEF/UFSM, 2005.

REZENDE, J. L. P.; OLIVEIRA A. D. **Análise econômica e social de projetos florestais**. Viçosa: UVF, 2001.

RIBEIRO, A. et al . Diametric distribution structure in an experimental candeia (*Eremanthus erythropappus* (DC.) MacLeish) plantation. **Ciência Florestal**, Santa Maria , v. 24, n. 4, p. 1055-1065, dec./2014. Available at: <http://www.scielo.br/scielo.php?script =sci_arttext&pid=S1980-50982014000401055&lng=en&nrm=iso>. Accessed on: 30/09/2018.

RODRIGUES, J. R. Roteiro para apresentação de Estudo de Impacto Ambiental - EIA and Relatório de Impacto Ambiental - RIMA. In: Verdum Roberto & Medeiros, Rosa M. V. (Org.). RIMA, Relatório de Impacto Ambiental: Legislação, elaboração e resultados, 4th ed. Porto Alegre: UFRGS, 2002.

ROSA, S.L. Passos metodológicos para o uso corrente de matriz de impactos na avaliação de projetos complexos. São Paulo: Octa Consultoria e Planejamento, Série Fascículos OCTA, n° 13, 1993.

SALOMON, M. J.; EDIN, O. **Análises de projetos**, 3. ed. Rio de Janeiro, 1967. 324 p.

SAS. A simple regression model with correction of heteroscedasticity. Cary: SAS Institute, 2004.

SAS. The SAS System for Window - release 8.02. Cary: SAS Institute, 2001.

SCHILLING, A. C. **Influence of pruning on the quality of wood from the first thinning of *Pinus elliottii* Engelm**. Dissertation (Master's Degree in Forestry Engineering) - Master's Degree in Forestry Engineering, UFSM, Santa Maria, 1996. 73 p.

SCHNEIDER, P. R. **Regression analysis applied to forestry engineering**. Santa Maria: UFSM, CEPEF, 1998. 236 p.

SCHNEIDER, P. R. **Introduction to forest management**. Santa Maria: UFSM, 1993.

SCHNEIDER, P. R.; FINGER, C.A. G.; HOPPE, J. M. Effect of pruning intensity on the production of *Pinus elliottii* Engelm., planted on poor soil, in the State of Rio Grande do Sul. **Ciência Florestal**, Santa Maria, v.9, n.1, 1999. p. 35-46.

SCHNEIDER, P.R. **Manejo Florestal**: planejamento da produção florestal. Santa Maria: UFSM, CCR, DCF, CEPEF, 2002.

SCHULTE, B. et al. **SouthPro - a computer program for managing uneven-aged loblolly pine stands**. Madison, WI: U.S. Department of Agriculture, Forest Service, Forest Products Laboratory, Gen. Tech. Rep. FPL-GTR-112, 1998. 47p.

SCOLFORO, J. R. S.; RIOS, M. S.; OLIVEIRA, A. D. de; MELLO, J. M. de; MAESTRI, R. Accuracy of thinning equations to represent the stem profile of Pinus elliottii. **Cerne**, v.4, n.1, 1998. p. 100-122.

SFB. Forest Production. **SNIF Bulletin**, Brazilian Forest Service, Brasilia, 2ed., v.2, 2016.

SILVEIRA, B.D.; FLORIANO, E.P.; NAKAJIMA, N.Y.; HOSOKAWA, R.T.; ROSSOT, N.C.; GRACIOLI, C.R. Relationship of morphometry and competition with the growth of *Trichilia claussenii* in a semideciduous forest fragment, RS. **FLORESTA**, Curitiba, PR, v. 45, n. 2, p.373-382, apr./jun, 2015.

SILVESTRE, R.; BONAZZA, M.; STANG, M.; LIMA, G.C.P.; KOEPSEL, D.A.; MARCO, F.T.; CIARNOSCHI, L.D.; SCARIOT, R.; MORÊS, D.F. Volumetric equations in *Pinus taeda* L. stands in the municipality of Lages-SC. **Nativa**, Sinop, v. 02, n. 01, p. 01-05, jan./mar. 2014.

SIT, V. **Catalog of curves for curve fitting**. Victoria, Canada, B.C.: Ministry of Forests, Forest Science Research Branch, Biometrics information handbook series, ISSN 1183-9759-no.4, 1994. 110 p.

SNIF. **Forest resources**: planted forests. National Forestry Information Service, 2018. Available at: http://www.florestal.gov.br/snif/recursos-florestais/as-florestas-plantadas. Accessed in: 2018.

SOARES, S.R. **EIA-RIMA/Licenciamento - Ciclo** de **vida / Ganhos ambientais / Avaliação de sistemas - Parte 10**. Universidade Federal de Santa Catarina, Programa de pós-graduação em engenharia ambiental, 2002-b. Available at: http://www.ens.ufsc.br/~soares. Accessed on: 23/Nov/2002.

SOARES, S.R. **Example of a matrix for assessing the significance of impacts**. Florianópolis: Federal University of Santa Catarina, Department of Sanitary and Environmental Engineering, 2002-b. Available at: <http://www.ens. ufsc.br/~soares> . Accessed on: 23/Jan/2003.

SOARES, S.R. **Environmental Indicators - Environmental Labeling**. Florianópolis: Federal University of Santa Catarina, Department of Sanitary and Environmental Engineering, 2002. Available at: <http://www.ens. ufsc.br/~soares> . Accessed on: 23/Jan/2003.

SOARES, S.R. **Environmental management system**. Florianópolis: Federal University of Santa Catarina, Department of Sanitary and Environmental Engineering, ENS 5125 - Environmental Management and Planning, Classes 06, 07 and 08, 2002-a. Available at: <http://www.ens. ufsc.br/~soares>. Accessed on: January 23, 2003.

SOARES, T. S. et al. Economic evaluation of *Eucalyptus* plantations submitted to thinning. **Revista Árvore**, v.27, n.4, p.481-486, Viçosa, jul-ago/2003.

SOARES, Thelma S. **Optimizing the use of wood in eucalyptus stands**. Viçosa, 2002. 60 f. Dissertation (Master's Degree in Forest Science) - Postgraduate Program in Forest Science, UFV, Viçosa, 2002-c.

STORCK, L.; LOPES, S. J. **Experimentation II**. Santa Maria: UFSM, CCR/Dep. Fitotecnia, 1998. 205 p.

TARGET. **Sustainable forest management** - NBR 14789: The principles, criteria and indicators for forest plantations. São Paulo: Target Engenharia e Consultoria, 2012. Available at: <https://www.target.com.br/produtossolucoes/solucoes/Solucoes.aspx?pp=27&c=2925>. Accessed on: 29/09/2013.

TOMÉ, M. **Inventory of forest resources**. Lisbon: Universidade Técnica de Lisboa / ISA / GIMREF, Textos pedagógicos, n. 1, 2002.

UBIALLI, J. A. Comparison of sampling methods and processes for phytosociological studies and stock estimates of an ecotonal forest in the northern region of Mato Grosso. Thesis (Doctorate) - UFPR/PPGEF, Curitiba, 2007. 241 f.

UFSM/SEMA-RS. Continuous forest inventory - RS. [Santa Maria]: UFSM, SEMA-RS, 2003. Available at: <http://coralx.ufsm.br/ifcrs/>. Accessed on: 07/11/2003.

VANCLAY, J. Review of Competition Indices: What have we learned, where should we use then and what additional research is required? - IN:IUFRO CENTENNIAL MEETING, Eberswalde - Berlin, 31 August - 6 September 1992.

VIALE, R. D. **Arborização urbana**: mileira barriguda (Ceiba speciosa). Pimenta Bueno, RO: City Hall/SEMMA, 2023. Available at: <https://pimentabueno.sedam.ro.gov.br/paineira-barriguda-ceiba-speciosa/>. Accessed on: 2024.

WARNER, A. Pruning. **Private Forests Tasmania**, [Hobart], Technical Information Sheet No. 20, March 1997.

WEST, P.W. **Tree and Forest Measurement**, 2 ed. Berlin/Heidelberg: Sringer-Verlag, 2009.

WILLIAMS, M. R. W. **Decision-making in forest management**, 2nd ed. Great Yarmouth: Research Studies Press/John Wiley & Sons, 1988.

WILSON, F.G. Numerical Expression of Stocking in Terms of Height. **Journal of Forestry**, Washington, v.44, n.10, p.758-761, October-1946.

WONNACOTT, T. H.; WONNACOTT, R. J. **Introduction to statistics.** Rio de Janeiro: Technical and Scientific Books, 1980. 589 p.

WYCOFF, W. R. et al. **User's guide to the stand prognosis model**. Ogden, UT: Intermountain Forest and Range Experiment Station, FS-USDA, GTR. INT-133, 1982. 112 p.

Anexo 1. Volumetric equation models (SILVESTRE et al, 2014)

Nº	Autor	Modelo
1	Burkhault I	$v = \beta 0 + DAP\beta 1 + ei$
2	Kopezky-Gehrhatdt I	$v = \beta 0 + \beta 1 \cdot DAP + ei$
3	Kopezky-Gehrhatdt II	$v = \beta 0 + \beta 1 \cdot DAP^2 + ei$
4	Dissescu-Meyer	$v = \beta 1 \cdot DAP + \beta 2 \cdot DAP^2 + ei$
5	HohenadlKrenn	$v = \beta 0 + \beta 1 \cdot DAP + \beta 2 \cdot DAP^2 + ei$
6	Bonetes I	$v = \beta 0 + \beta 1 \cdot (1/DAP) + \beta 2 \cdot (1/DAP)^2 + ei$
7	Husch-1963	$Log(v) = \beta 0 + \beta 1 \cdot log(DAP) + ei$
8	Omerod MI	$v = \beta 0 + \beta 1 \cdot DAP \cdot h^2 + \beta 2 \cdot DAP^2 + ei$
9	Omerod MII	$v = \beta 0 + \beta 1 \cdot DAP^2 \cdot h + \beta 2 \cdot (1/\sqrt{DAP}) + ei$
10	Spurr FF constante	$v = \beta 1 \cdot DAP^2 \cdot h + ei$
11	Spurr Var combinada	$v = \beta 0 + \beta 1 \cdot DAP \cdot h + ei$
12	Spurr Var combinada II	$v = \beta 0 + \beta 1 \cdot DAP^2 \cdot h + ei$
13	Bonetes II	$v = \beta 0 + \beta 1 \cdot DAP + \beta 2 \cdot h + ei$
14	Bonetes III	$v = \beta 0 + \beta 1 \cdot DAP^2 + \beta 2 \cdot h + ei$
15	Bonetes IV	$v = \beta 0 + \beta 1 \cdot DAP^2 + \beta 2 \cdot h^2 + ei$
16	Stoate australiana	$v = \beta 0 + \beta 1 \cdot DAP^2 + \beta 2 \cdot DAP^2 \cdot h + \beta 3 \cdot h + ei$
17	Burkhaut II	$v = \beta 0 + \beta 1 \cdot \sqrt{DAP} \cdot h + \beta 2 \cdot (1/DAP) + \beta 3 \cdot DAP^2 + ei$
18	Burkhaut III	$v = \beta 0 + \beta 1 \cdot h^2 + \beta 2 \cdot DAP^2 + \beta 3 \cdot (1/\sqrt{DAP}) + ei$
19	Meyer modificada I	$v = \beta 0 + \beta 1 \cdot DAP \cdot h + \beta 2 \cdot DAP^2 \cdot h^2 + \beta 3 \cdot DAP^3 \cdot h^3 + \beta 4 \cdot DAP4 \cdot h4 + ei$
20	Meyer modificada II	$v = \beta 0 + \beta 1 \cdot DAP + \beta 2 \cdot DAP^2 + \beta 3 \cdot DAP \cdot h + \beta 4 \cdot DAP^2 \cdot h + ei$
21	Meyer compreensiva	$v = \beta 0 + \beta 1 \cdot DAP + \beta 2 \cdot DAP^2 + \beta 3 \cdot DAP \cdot h + \beta 4 \cdot DAP^2 \cdot h + \beta 5 \cdot h + ei$
22	Näslund-spruce I	$v = \beta 0 + \beta 1 \cdot DAP^2 + \beta 2 \cdot DAP^2 \cdot h + \beta 3 \cdot DAP \cdot h^2 + \beta 4 \cdot h + ei$
23	Näslund-spruce II	$v = \beta 0 + \beta 1 \cdot DAP^2 + \beta 2 \cdot ln(DAP^2 \cdot h) + \beta 3 \cdot DAP \cdot h^2 + \beta 4 \cdot h^2 + ei$
24	Clutter	$v = e\beta 0 + \beta 1 \cdot h + \beta 2 \cdot log(DAP) + ei$
25	Clutter II	$v = e\beta 0 + \beta 1 \cdot log^2(h) + \beta 2 \cdot log(DAP) + ei$
26	Takata	$v = DAP \cdot h/(\beta 0 + \beta 1 \cdot (1/DAP)) + ei$
27	Spurr-Var combinada III	$ln(v) = \beta 0 + \beta 1 \cdot ln(DAP^2 \cdot h) + ei$
28	I.B.W. Alemanha	$ln(v) = \beta 0 + \beta 1 \cdot ln(DAP) + \beta 2 \cdot ln^2(DAP) + \beta 3 \cdot ln(h) + \beta 4 \cdot ln^2(h) + ei$
29	Schumacher Hall não linear	$v = \beta 0 \cdot (DAP\beta 1) \cdot (h\beta 2) + ei$
30	Schumacher Hall	$ln(v) = \beta 0 + \beta 1 \cdot ln(DAP) + \beta 2 \cdot ln(h) + ei$
31	Stoate	$Log(v) = \beta 0 + \beta 1 \cdot log^2(h) + \beta 2 \cdot log^2(DAP) + \beta 3 \cdot DAP^2 + ei$
32	Prodan II	$Log(v) = \beta 0 + \beta 1 \cdot log(d) + \beta 2 \cdot log^2(DAP) + \beta 3 \cdot log(h) + \beta 4 \cdot log^2(h) + ei$
33	Prodan I	$\sqrt{v} = \beta 0 + \beta 1 \cdot DAP \cdot h + \beta 2 \cdot DAP^2 \cdot h^2 + \beta 3 \cdot DAP^3/h^3 + \beta 4 \cdot DAP4/h4 + \beta 5 \cdot DAP5 \cdot h5 + \beta 6 \cdot DAP6 \cdot h6 + ei$

Em que: v = volume; DAP = diâmetro à altura do peito (1,30 m); h = altura total; $\beta 0$, $\beta 1$, $\beta 2$, $\beta 3$, $\beta 4$, $\beta 5$, $\beta 6$ = coeficientes; ln = logaritmo neperiano; e = exponencial; log = logaritmo base 10; ei = erro aleatório.

Anexo 2. Trunk tapering models (MÜLLER, 2004)

Nº	Modelos de Funções de Forma	Autores
01	$\frac{d_i}{d} = b_0 + b_1.\left(\frac{h_i}{h - 1{,}30}\right)$	Munro (1966)
02	$\frac{d_i^2}{d^2} = b_0 - b_1.\left(\frac{h_i}{h - 1{,}30}\right) + b_2.\left(\frac{h_i}{h - 1{,}30}\right)^2$	
03	$\frac{d_i^2}{d^2} = b_1.\left(\frac{h_i}{h - 1{,}00}\right) + b_2.\left(\frac{h_i^2}{h^2 - 1{,}00}\right)$	Kozak *et al.* (1969)
04	$\frac{d_i^2}{d^2} = b_0 + b_1.\left(\frac{h_i}{h - 1{,}00}\right) + b_2.\left(\frac{h_i^2}{h^2 - 1{,}00}\right)$	Kozak *et al.* (1969) Modificada
05	$ln\left(\frac{d_i}{d}\right) = b_0 + b_1.ln\left(\frac{1}{h_i + 1{,}30}\right)$	Ormerad (1973)
06	$\frac{d_i}{d} = b_0 + b_1.\left(\frac{1}{h_i + 1{,}30}\right) + b_2.\left(\frac{h}{h_i + 1{,}30}\right) + b_3.\left(\frac{1}{h_i + 1{,}30}\right).(h + h_i + 1{,}30)$	Bennett & Swindel (1972) Modificada
07	$\frac{d_i}{d} = b_0 + b_1.\left(\frac{1}{h_i + 1{,}30}\right)^{1/2}$	Gray (1956) Modificada
08	$ln\left(\frac{d_i}{d}\right) = b_0 + b_1.ln\left(\frac{h - h_i}{h}\right)$	
09	$ln\left(\frac{d_i}{d}\right) = b_0 + b_1.ln\left(\frac{h - h_i}{h}\right) + b_2.ln^2\left(\frac{h - h_i}{h}\right)$	Silva & Sterba (1975)
10	$ln\left(\frac{d_i}{d}\right) = b_0 + b_1.ln\left(\frac{h}{h + h_i}\right)$	
11	$\frac{d_i}{d} = b_0 + b_1.\left(\frac{h}{h + h_i}\right) + b_2.\left(\frac{h}{h + h_i}\right)^2 + b_3.\left(\frac{h}{h + h_i}\right)^3$	
12	$\frac{d_i}{d} = b_0 + b_1.\left(\frac{h}{h + h_i}\right) + b_2.\left(\frac{h}{h + h_i}\right)^2 + b_3.\left(\frac{h}{h + h_i}\right)^3 + b_4.\left(\frac{h}{h + h_i}\right)^4$	
13	$\frac{d_i}{d} = b_0 + b_1.\left(\frac{h}{h + h_i}\right) + b_2.\left(\frac{h}{h + h_i}\right)^2 + b_3.\left(\frac{h}{h + h_i}\right)^3 + b_4.\left(\frac{h}{h + h_i}\right)^4 + b_5.\left(\frac{h}{h + h_i}\right)^5$	
14	$\frac{d_i^2}{d^2} = b_0 + b_1.\left(\frac{h_i}{h}\right)^2$	Munro (1968)
15	$\frac{d_i^2}{d^2} = b_0 + b_1.\left(\frac{h_i}{h}\right) + b_2.\left(\frac{h_i}{h}\right)^2$	Munro (1968)
16	$\frac{d_i}{d} = b_0 + b_1.\left(\frac{h_i}{h}\right) + b_2.\left(\frac{h_i}{h}\right)^2 + b_3.\left(\frac{h_i}{h}\right)^3$	
17	$\frac{d_i}{d} = b_0 + b_1.\left(\frac{h_i}{h}\right) + b_2.\left(\frac{h_i}{h}\right)^2 + b_3.\left(\frac{h_i}{h}\right)^3 + b_4.\left(\frac{h_i}{h}\right)^4$	
18	$\frac{d_i}{d} = b_0 + b_1.\left(\frac{h_i}{h}\right) + b_2.\left(\frac{h_i}{h}\right)^2 + b_3.\left(\frac{h_i}{h}\right)^3 + b_4.\left(\frac{h_i}{h}\right)^4 + b_5.\left(\frac{h_i}{h}\right)^5$	
19	$f_K(X) = A_K(X - X_K)^3 + B_K(X - X_K)^2 + C_K(X - X_K) + D_K$	

Onde: d_i = Diâmetro a uma altura relativa h_i, em centímetros; d = Diâmetro à altura do peito (1,30 m), em centímetros; h_i = Altura na posição i, em metros; h = Altura total da árvore, em metros; b_0, b_1, b_2, b_3, b_4, b_5, A_k, B_k, C_k e D_k = Coeficientes; ln = Logaritmo natural.

Anexo 3. Competition indices (CONTRERAS et al, 2011).

Index	Sources	Equations
Distance-independent competition indices		
CI_1	Reineke (1933)	$10^{(\log N + 1.605 \times \log dg - 1.605)}$
CI_2	Referred as canopy closure in Rivas et al. (2005)	$(\sum_{i=1}^{n}(\pi \times cw_i^2)/4)/S$
CI_3	Wykoff et al. (1982)	$(\sum_{j=1}^{n}(\pi \times d_{maxj}^2)/4) = BAL$
CI_4	Schröder and Gadow (1999)	$(1 - [1 - (BAL/G)])/RS$
Distance-dependent competition indices		
CI_5	Number of neighbors	n
CI_6	Gerrard (1969)	$\sum_{i=1}^{n} O_i/Z$
CI_7	Bella (1971)	$\sum_{i=1}^{n}(O_i \times d_i)/(Z \times d)$
CI_8	Hegyi (1974)	$\sum_{i=1}^{n} d_i/(d \times dist_i)$
CI_9	Braathe (1980), cited in Pukkala and Kolström (1987)	$\sum_{i=1}^{n} h_i/(h \times dist_i)$
CI_{10}	Rouvinen and Kuuluvainen (1997)	$\sum_{i=1}^{n} \arctan(d_i/dist_i)$
CI_{11}	Rouvinen and Kuuluvainen (1997)	$\sum_{i=1}^{n}(d_i/d) \times \arctan(d_i/dist_i)$
CI_{12}	Rouvinen and Kuuluvainen (1997)	$\sum_{i=1}^{n} \arctan(h_i/dist_i)$
CI_{13}	Rouvinen and Kuuluvainen (1997)	$\sum_{i=1}^{n}(h_i/h) \times \arctan(h_i/dist_i)$
Light values		
CI_{14}	Sky's hemisphere	$(\sum_{i=1}^{2160} B_i)/2160$
CI_{15}	Sky's hemisphere/sun position	$(\sum_{i=1}^{1080} B_{Ni}) + (\sum_{i=1}^{1080} 2 \times B_{Si})/3180$
CI_{16}	Sun position	$(\sum_{i=1}^{63} B_i)/108$

Where: N = trees per ha in the plot; dg = quadratic mean diameter (cm); n = number of neighbors within the 11 m radius competition area; cwi = crown width of the neighboring i-tree (m); S = plot area (m2); BAL = basal area of neighboring trees larger than the central tree (m^2/ha); G = total basal area of trees within 11 m (m^2/ha); RS = relative spacing index of the plot; Oi Area of the overlapping zone of influence between the i-neighboring tree and the central tree (m^2); Z = area of the zone of influence of the central tree (m^2); di = DBH of the next i-tree (cm); d = DBH of the central tree (cm); disti = horizontal distance from the neighboring i-tree to the central tree (m); hi = height of the neighboring i-tree (m); h = height of the central tree (m); Bi = Binary variable that is 1 if light ray i is blocked, or 0 otherwise; BNi = Binary variable associated with light rays oriented to the north (azimuths from 270 to 90), 1 if blocked and 0 otherwise; BSi = Binary variable associated with light rays oriented to the south (azimuths from 90 to 270), 1 if blocked and 0 otherwise.

Anexo 4. Competition indices (MALEKI et al, 2015).

Index	Sources	Equations
Non-spatially explicit indices		
BA-g $BA\text{-}g_j$	Steneker & Jarvis (1963)	$\sum_{j\neq i}^{n}(g_j)/S$
BAL	Wykoff *et al.* (1982)	$\sum_{d_i<d_j}^{n}(g_j)/S$
Sdr	Lorimer (1983)	$\left(\left(\sum_{j\neq i}^{n}d_j\right)/d_i\right)/S$
dr_g	Hamilton (1986)	d_i/d_g
BAr	Corrona & Ferrara (1989)	$\left(\left(\sum_{j\neq i}^{n}g_j\right)/g_i\right)/S$
$BALr$	Vanclay (1991)	$\sum_{j\neq i}^{n}(g_{j,dj>d_i})/G$
$BALMOD$	Schröder & Gadow (1999)	$\left[\left(\sum_{d_i<d_j}^{n}(g_j)/G\right)\right]/RS$
Spatially explicit indices		
SI	Staebler (1951)	$\sum_{j\neq i}^{n}l_{ij}$
SOr	Gerrard (1969)	$\sum_{j\neq i}^{n}(O_{ij}/CZ)$
$SOdr$	Bella (1971)	$\sum_{j\neq i}^{n}((O_{ij}\cdot d_j)/(CZ\cdot d_i))$
$SBAr$	Daniels et al. (1986)	$\left((d_i^2\cdot N_c)/\sum_{j\neq i}^{n}d_j^2\right)$
Heg	Hegyi (1974)	$\sum_{j\neq i}^{n}(d_j/(d_i\cdot l_{ij}))$
$SAng1$	Lin (1974)	$2\sum_{j\neq i}^{n}\arctan(d_j/2l_{ij})$
$SAng2$	Rouvinen & Kuuluvainen (1977)	$\sum_{j\neq i}^{n}\arctan(d_j/l_{ij})$
$SdrAng$	Rouvinen & Kuuluvainen (1977)	$\sum_{j\neq i}^{n}((d_j/d_i)\cdot\arctan(d_j/l_{ij}))$
$Almdg$	Alemdag (1978)	$\sum_{j\neq i}^{n}\left\{\pi\left[(l_{ij}\cdot d_i)/(d_i+d_j)\right]^2(d_j/l_{ij})/\sum(d_j/l_{ij})\right\}$
$Sdrl1$	Lorimer (1983)	$\sum_{j\neq i}^{n}\left((d_j/d_i)/\sqrt{l_{ij}/CZR}\right)$
$Sdrl2$	Martin & Ek (1984)	$\sum_{j\neq i}^{n}(d_j/d_i)\cdot\exp((16\cdot l_{ij})/(d_i+d_j))$

Where: Nc = number of competing trees; di = d of central tree (cm); dj = d of competing tree (cm); dg = plot diameter weighted by basal area (cm); lij = distance between competitor j and central tree i (m); G = basal area of trees within plot (m²/ha); gi = basal area of central tree i (m²/ha); gj = basal area of competitor tree j (m²/ha); CZR = radius of the zone of influence (m); RS = relative spacing index of the plot; Oij = crown overlap between neighboring tree j and central tree i (m2); CZ = the area of the zone of influence (m²); S = plot area (ha).

MIX
Papier aus verantwortungsvollen Quellen
Paper from responsible sources
FSC® C105338

Printed by Books on Demand GmbH, Norderstedt / Germany